Building AI Applications with Microsoft Semantic Kernel

Easily integrate generative AI capabilities and copilot experiences into your applications

Lucas A. Meyer

Building AI Applications with Microsoft Semantic Kernel

Group Product Manager: Niranjan Naikwadi

Publishing Product Manager: Tejashwini R

Book Project Manager: Neil D'mello

Senior Editor: Mark D'Souza

Technical Editor: K Bimala Singha

Copy Editor: Safis Editing

Proofreader: Mark D'Souza

Indexer: Pratik Shirodkar

Production Designer: Shankar Kalbhor

DevRel Marketing Coordinator: Vinishka Kalra

First published: June 2024

Production reference: 1060624

Published by

Packt Publishing Ltd.

Grosvenor House

11 St Paul's Square

Birmingham

B3 1RB, UK

ISBN 978-1-83546-370-3

www.packtpub.com

Contributors

About the author

Lucas A. Meyer is a financial economist and computer scientist with an MBA and an M.Sc. in finance from the University of Washington in Seattle. Lucas works as a principal research scientist at the Microsoft AI for Good Lab, where he works on the economics of AI and using large language models to combat disinformation and improve economic outcomes. Prior to that, Lucas worked for several years in finance, pioneering several uses of natural language processing for financial analysis and automation. Lucas is passionate about AI education and is a Top Voice on LinkedIn, where he posts about new developments in the field.

To my wife, Ilanah, and my children, Thomas, Eric, and Caroline: you make it easy and worthwhile for me to face any challenge. To my mom, Ana Maria, thank you for teaching me that education is the best investment. To my colleagues at the AI for Good Lab: I learn so much from you. To the online communities of LinkedIn and Threads, thanks for all you taught me and all the support.

About the reviewer

Lucas Puskaric has driven billions of dollars in revenue as a staff software engineer and tech lead, specializing in crafting creative customer-focused solutions across the entire stack. Over his career, he has worked at every type of company from non-profit to Fortune 25. After helping vaccinate tens of millions of people, he founded the software company Interweave. When LLMs arrived on the scene, he was an early adopter who released several AI apps. He's known for his presence on Threads, where he often shares technical knowledge and projects he's working on. Connect with him at @lucaspuskaric on all socials.

Shoutout to the love of my life, Alex, and our four cats: Reggie, Sonny, Nori, and Figgy. Also, thanks to Threads for introducing me to Lucas Meyer and Packt for giving me the opportunity to review this masterpiece.

Table of Contents

Part 1: Introduction to Generative AI and Microsoft Semantic Kernel

1

2

Creating Better Prompts   37

Part 2: Creating AI Applications with Semantic Kernel

3

Extending Semantic Kernel   59

4

Performing Complex Actions by Chaining Functions 95

5

Programming with Planners 127

6

Adding Memories to Your AI Application 149

Part 3: Real-World Use Cases

7

Real-World Use Case – Retrieval-Augmented Generation   171

8

Real-World Use Case – Making Your Application Available on ChatGPT   199

Preface

Artificial intelligence is experiencing unprecedented growth, with new models emerging daily. With over 20 years in the technology sector, I can attest that the pace of innovation has never been this fast. This brings not only opportunities but also considerable change. Navigating these changes can be challenging and costly, as you may invest a lot of time learning a new technology that might become obsolete.

Enter Microsoft Semantic Kernel – a framework that reduces these risks by enabling access to various AI services through popular programming languages. This framework spares you the details of grappling with constantly evolving APIs. By learning Microsoft Semantic Kernel, you can write code at the framework level, and the framework will call the underlying models for you. This allows you to focus on core concepts instead of the details of each model.

One of the key benefits of Semantic Kernel is its ability to use different AI services. For example, code initially targeting the OpenAI GPT platform can be switched to Google Gemini, often without any modifications. This flexibility makes it easier to integrate AI into applications and to make minimal modifications to them when change inevitably happens.

Moreover, Semantic Kernel makes AI accessible to enterprise programming languages. While Python has long dominated the AI landscape, many enterprise applications rely on C# or Java. Recognizing this, Semantic Kernel not only supports Python but also elevates C# to a first-class AI language. Java support is currently in its beta stages and is expected to launch fully in 2024.

Whether you're a solo developer or part of a larger enterprise, the demand to add AI functionality to applications is inevitable. This book was created to equip you with the necessary skills to implement AI quickly and effectively, ensuring you are well-prepared to meet this growing demand.

Who this book is for

The three main personas who are the target audience of this book are as follows:

- **Developers** who want to add AI to their applications without adding the complexity of connecting to several different services and maintaining the connections after each release
- **Technical program managers** who can write some code and want to quickly prototype AI functionality without having to learn details
- **Applied researchers and data scientists** who want to focus their time on solving business problems quickly, instead of working on the ever-changing ways of connecting and using AI services

What this book covers

Chapter 1, Introducing Microsoft Semantic Kernel, introduces several AI concepts and gives a tour of what Semantic Kernel can help you achieve, showing you how to connect to an AI service and use it to achieve a goal.

Chapter 2, Creating Better Prompts, teaches you several techniques on how to interact better with AI, improving the chances that you will get a good result on your first try, using a concept called **prompt engineering**.

Chapter 3, Extending Semantic Kernel, teaches you how to add functionality to Semantic Kernel, by adding **native functions** and **semantic functions** that can later be reused by you, as a developer, or your user to achieve their goals.

Chapter 4, Performing Complex Actions by Chaining Functions, shows you how to use several functions of a kernel in sequence, making programming complex actions a lot easier.

Chapter 5, Programming with Planners, explores how Semantic Kernel can receive a request in natural language and automatically decide which functions to call to achieve an objective, allowing users of your application to perform functions that you did not have to write code for.

Chapter 6, Adding Memories to Your AI Application, examines how to add external knowledge to the AI models used by Semantic Kernel, making it easier for AI models to remember recent conversations and personalizations.

Chapter 7, Real-World Use Case – Retrieval-Augmented Generation, shows how to add a large amount of data to AI models, allowing them to efficiently use information that they have not been trained on, including very recent and private data.

Chapter 8, Real-World Use Case – Making Your Application Available on ChatGPT, shows how to publish the application you wrote with Microsoft Semantic Kernel on OpenAI's GPT store, making it instantly available to millions of users.

To get the most out of this book

Readers will have to be familiar with programming in either Python or C#, and familiar with creating and connecting to AI services on the cloud.

Software/hardware covered in the book	Operating system requirements
Python 3.11	Windows, macOS, or Linux
.NET 8	
OpenAI GPT-3.5 and GPT-4	

For *Chapter 7*, you will need to create an index in Azure AI search. There is a free tier. For *Chapter 8*, to create a GPT to share with others, you will need a ChatGPT subscription.

If you are using the digital version of this book, we advise you to type the code yourself or access the code from the book's GitHub repository (a link is available in the next section). Doing so will help you avoid any potential errors related to the copying and pasting of code.

Download the example code files

You can download the example code files for this book from GitHub at `https://github.com/ PacktPublishing/Building-AI-Applications-with-Microsoft-Semantic- Kernel`. If there's an update to the code, it will be updated in the GitHub repository.

We also have other code bundles from our rich catalog of books and videos available at `https:// github.com/PacktPublishing/`. Check them out!

Conventions used

There are a number of text conventions used throughout this book.

`Code in text`: Indicates code words in text, database table names, folder names, filenames, file extensions, pathnames, dummy URLs, user input, and Twitter handles. Here is an example: "Since our prompt will be a new function and have multiple parameters, we will also need to create a new `config.json` file."

A block of code is set as follows:

```
    response = await kernel.invoke(pe_plugin["chain_of_thought"],
KernelArguments(problem = problem, input = solve_steps))
    print(f"\n\nFinal answer: {str(response)}\n\n")
```

When we wish to draw your attention to a particular part of a code block, the relevant lines or items are set in bold:

```
[default]
exten => s,1,Dial(Zap/1|30)
exten => s,2,Voicemail(u100)
exten => s,102,Voicemail(b100)
exten => i,1,Voicemail(s0)
```

Any command-line input or output is written as follows:

```
dotnet add package Microsoft.SemanticKernel.
s.Handlebars --version 1.0.1-preview
```

Bold: Indicates a new term, an important word, or words that you see onscreen. For instance, words in menus or dialog boxes appear in **bold**. Here is an example: "S Once these configurations are done, click **Review + create** and your web application will be deployed in a few minutes."

> **Tips or important notes**
> Appear like this.

Get in touch

Feedback from our readers is always welcome.

General feedback: If you have questions about any aspect of this book, email us at customercare@ packtpub.com and mention the book title in the subject of your message.

Errata: Although we have taken every care to ensure the accuracy of our content, mistakes do happen. If you have found a mistake in this book, we would be grateful if you would report this to us. Please visit www.packtpub.com/support/errata and fill in the form.

Piracy: If you come across any illegal copies of our works in any form on the internet, we would be grateful if you would provide us with the location address or website name. Please contact us at copyright@packt.com with a link to the material.

If you are interested in becoming an author: If there is a topic that you have expertise in and you are interested in either writing or contributing to a book, please visit authors.packtpub.com.

Share Your Thoughts

Once you've read *Building AI Applications with Microsoft Semantic Kernel*, we'd love to hear your thoughts! Scan the QR code below to go straight to the Amazon review page for this book and share your feedback.

https://packt.link/r/1-835-46370-3

Your review is important to us and the tech community and will help us make sure we're delivering excellent quality content.

Download a free PDF copy of this book

Thanks for purchasing this book!

Do you like to read on the go but are unable to carry your print books everywhere?

Is your eBook purchase not compatible with the device of your choice?

Don't worry, now with every Packt book you get a DRM-free PDF version of that book at no cost.

Read anywhere, any place, on any device. Search, copy, and paste code from your favorite technical books directly into your application.

The perks don't stop there, you can get exclusive access to discounts, newsletters, and great free content in your inbox daily

Follow these simple steps to get the benefits:

1. Scan the QR code or visit the link below

https://packt.link/free-ebook/9781835463703

2. Submit your proof of purchase
3. That's it! We'll send your free PDF and other benefits to your email directly

Part 1: Introduction to Generative AI and Microsoft Semantic Kernel

In this part, you will get an overview of generative AI and how to use it with Microsoft Semantic Kernel. In addition to this, you will also learn best practices for prompting that can be useful not only when you're using Semantic Kernel, but also when you're interacting with AI on your own.

This part includes the following chapters:

Chapter 1, Introducing Microsoft Semantic Kernel

Chapter 2, Creating Better Prompts

1

Introducing Microsoft Semantic Kernel

The **generative artificial intelligence** (**GenAI**) space is evolving quickly, with dozens of new products and services being launched weekly; it is becoming hard for developers to keep up with the ever-changing features and **application programming interfaces** (**APIs**) for each of the services. In this book, you will learn about **Microsoft Semantic Kernel**, an API that will make it a lot easier for you to use GenAI as a developer, making your code shorter, simpler, and more maintainable. Microsoft Semantic Kernel will allow you, as a developer, to use a single interface to connect with several different GenAI providers. Microsoft used Semantic Kernel to develop its copilots, such as Microsoft 365 Copilot.

Billions of people already use GenAI as consumers, and you are probably one of them. We will start this chapter by showing some examples of what you can do with GenAI as a consumer. Then, you will learn how you can start using GenAI as a developer to add AI services to your own applications.

In this chapter, you will learn the differences between using GenAI as a user and as a developer and how to create and run a simple end-to-end request with Microsoft Semantic Kernel. This will help you see how powerful and simple Semantic Kernel is and will serve as a framework for all further chapters. It will enable you to begin integrating AI into your own apps right away.

In this chapter, we'll be covering the following topics:

- Understanding the basic use of a generative AI application like ChatGPT
- Installing Microsoft Semantic Kernel
- Configuring Semantic Kernel to interact with AI services
- Running a simple task using Semantic Kernel

Technical requirements

To complete this chapter, you will need to have a recent, supported version of your preferred Python or C# development environment:

- For Python, the minimum supported version is Python 3.10, and the recommended version is Python 3.11

- For C#, the minimum supported version is .NET 8

> **Important note**
>
> The examples are presented in C# and Python, and you can choose to only read the examples of your preferred language. Occasionally, a feature is available in only one of the languages. In such cases, we provide an alternative in the other language for how to achieve the same objectives.

In this chapter, we will call OpenAI services. Given the amount that companies spend on training these large language models (LLMs), it's no surprise that using these services is not free. You will need an **OpenAI API** key, obtained either directly through **OpenAI** or **Microsoft**, via the **Azure OpenAI** service.

> **Important: Using the OpenAI services is not free**
>
> The examples that we will run throughout this book will call the OpenAI API. These calls require a paid subscription, and each call will incur a cost. The costs are usually small per request (for example, GPT-4 costs up to $0.12 per 1,000 tokens), but they can add up. In addition, note that different models have different prices, with GPT-3.5 being 30 times less expensive per token than GPT-4.
>
> OpenAI pricing information can be found here: `https://openai.com/pricing`
>
> Azure OpenAI pricing information can be found here: `https://azure.microsoft.com/en-us/pricing/details/cognitive-services/openai-service/`

If you use .NET, the code for this chapter is at `https://github.com/PacktPublishing/Building-AI-Applications-with-Microsoft-Semantic-Kernel/tree/main/dotnet/ch1`.

If you use Python, the code for this chapter is at `https://github.com/PacktPublishing/Building-AI-Applications-with-Microsoft-Semantic-Kernel/tree/main/python/ch1`.

You can install the required packages by going to the GitHub repository and using the following: `pip install -r requirements.txt`.

Obtaining an OpenAI API key

1. Go to the OpenAI Platform website (`https://platform.openai.com`).

2. Sign up for a new account or sign in with an existing account. You can use your email or an existing Microsoft, Google, or Apple account.

3. On the left sidebar menu, select **API keys**.

4. On the **Project API keys** screen, click the button labeled + **Create new secret key** (optionally, give it a name).

> **Important**
> You have to copy and save the key immediately. It will disappear as soon as you click **Done**. If you didn't copy the key or if you lost it, you need to generate a new one. There's no cost to generate a new key. Remember to delete old keys.

Obtaining an Azure OpenAI API key

Currently, you need to submit an application to obtain access to the Azure OpenAI Service. To apply for access, you need to complete a form at `https://aka.ms/oai/access`.

The instructions to obtain an Azure OpenAI API key are available at `https://learn.microsoft.com/en-us/azure/ai-services/openai/how-to/create-resource`.

Generative AI and how to use it

Generative AI refers to a subset of artificial intelligence programs that are capable of creating content that is similar to what humans can produce. These systems use training from very large datasets to learn their patterns, styles, and structures. Then, they can generate entirely new content, such as synthesized images, music, and text.

Using GenAI as consumer or end-user is very easy, and as a technical person, you probably have already done it. There are many consumer-facing AI products. The most famous is OpenAI's **ChatGPT**, but there are many others that have hundreds of millions of users every day, such as Microsoft Copilot, Google Gemini (formerly Bard), and Midjourney. As of October 2023, Meta, the parent company of Facebook, WhatsApp, and Instagram, is making GenAI services available to all its users, increasing the number of GenAI daily users to billions.

While the concept of GenAI has existed for a while, it gained a lot of users with the release of OpenAI's ChatGPT in November of 2022. The initial version of ChatGPT was based on a model called **generative pre-trained transformer** (**GPT**) version 3.5. That version was a lot better than earlier versions in the task of mimicking human-like writing. In addition, OpenAI made it easy to use by adding a chatbot-like interface and making it available to the general public. This interface is called ChatGPT.

With ChatGPT, users can easily initiate tasks in their own words. At its launch, ChatGPT was the product with the fastest adoption rate in history.

The GenAI concept was further popularized with the release of Midjourney, an application that allows for the generation of high-quality images from prompts submitted through Discord, a popular chat application, and Microsoft Copilot, a free web application that can generate text by using GPT-4 (the newest version of OpenAI's GPT) and images by using an OpenAI model called DALL-E 3.

In the upcoming subsections, we will discuss text and image generation using GenAI applications and explain the differences between generating them using applications such as ChatGPT and with an API as a developer.

Text generation models

The initial use cases for GenAI were to generate text based on a simple instruction called a **prompt**.

The technology used behind most text-based GenAI products is called a **transformer**, and it was introduced in the paper *Attention is All you Need* [1] in 2017. The transformer immensely improved the quality of the text being generated, and in just a few years, the text looked very similar to human-generated text. The transformer greatly improved the ability of AI to guess masked words in a phrase after being trained on a large number of documents (a **corpus**). Models trained on very large corpuses are called **large language models (LLMs)**.

If LLMs are given a phrase such as "*I went to the fast-food restaurant to <X>*," they can generate good options for X, such as "*eat*." Applying the transformer repeatedly can generate coherent phrases and even stories. The next iteration could be "*I went to the fast-food restaurant to eat <X>*," returning "*a*," and then "*I went to the fast-food restaurant to eat a <X>*," could return "*burger*," forming the full phrase "*I went to the fast-food restaurant to eat a burger*."

The performance of an LLM model depends on the number of parameters, which is roughly proportional to how many comparisons a model can make at once, the context window, the maximum size of the text that can be handled at once, and the data used to train the model, which is usually kept secret by the companies that create LLMs.

The GPT is a model created by OpenAI that uses transformers and is good at generating text. There are many versions of GPT:

GPT-1, launched in February 2018, had 120 million parameters and a **context window** of 512 tokens.

GPT-2 was launched in February 2019, with the number of parameters increasing to 1.5 billion and the context window increasing to 1,024 tokens. Up to this point, while they sometimes produced interesting results, these models were mostly used by academics.

This changed with GPT-3, launched in June 2020, which had several sizes: small, medium, large, and extra-large. Extra-large had 175 billion parameters and a 2,048 token context window. The generated text was, in most cases, hard to distinguish from human-generated text. OpenAI followed it with GPT-3.5, released in November 2022, with still 175 billion parameters and a context window of 4,096 tokens (now expanded to 16,384 tokens), and launched a user interface named ChatGPT.

ChatGPT is a web and mobile application that uses the GPT models in the background and allows users to submit prompts to the GPT models and get responses online. It was launched together with GPT-3.5, and at the time, it was the consumer product with the fastest adoption rate, reaching 100 million users in less than two months.

In February 2023, Microsoft released Bing Chat, which also uses OpenAI's GPT models in the back end, further popularizing the usage of transformer models and AI. Recently, Microsoft renamed it to Microsoft Copilot.

Just a month later, in March 2023, OpenAI released the GPT-4 model, which was quickly incorporated into the backend of consumer products such as ChatGPT and Bing.

Not all details about the GPT-4 model have been released to the public. It's known that it has a context window of up to 32,768 tokens; however, its number of parameters has not been made public, but it has been estimated at 1.8 trillion.

The GPT-4 model is notably good at human-like tasks related to text generation. A benchmark test shown in the GPT-4 technical report academic paper [2] shows the performance of GPT-3.5 and GPT-4 in taking exams. GPT-4 could pass many high-school and college level exams. You can read the paper at `https://doi.org/10.48550/arXiv.2303.08774`.

Understanding the difference between applications and models

Most people, including you, have likely used a GenAI application, such as ChatGPT, Microsoft Copilot, Bing Image Creator, Bard (now named Gemini), or Midjourney. These applications use GenAI models in their backend, but they also add a user interface and configurations that restrict and control the output of the models.

When you are developing your own application, you will need to do these things by yourself. You may not yet realize how much is carried out behind the scenes by applications such as Bing and ChatGPT.

When you submit a prompt to an application, the application may add several additional instructions to the prompt you submitted. The most typical things added are instructions to restrict some types of output, for example: "*your reply should contain no curse words.*" For example, when you submit the prompt "*Tell me a joke*" to an application like ChatGPT, it may modify your prompt to "*Tell me a joke. Your reply should contain no curse words*" and pass that modified prompt to the model.

The application may also add a summary of the questions that you have already submitted and the answers that were already given. For example, if you ask, "*How warm is Rio de Janeiro, Brazil, in the summer?*," the answer may be, "*Rio de Janeiro is typically between 90 and 100 degrees Fahrenheit (30-40 degrees Celsius) in the summer.*" If you then ask the question, "*How long is the flight from New York to there?*," an application such as ChatGPT will not submit "*How long is the flight from New York to there?*" directly to the model because the answer would be something like "*I don't know what you mean by 'there'.*"

A straightforward way to address this problem is to save everything that the user entered and all the answers that were provided and re-submit them with every new prompt. For example, when the user submits "*How long is the flight from New York to there?*" after asking about the temperature, the application prepends the earlier questions and answers to the prompt, and what is actually submitted to the model is: "*How warm is Rio de Janeiro, Brazil, in the summer? Rio de Janeiro is typically between 90 and 100 degrees Fahrenheit (30-40 degrees Celsius) in the summer. How long is the flight from New York to there?*" Now, the model knows that "*there*" means "*Rio de Janeiro,*" and the answer will be something like "*Approximately 10 hours.*"

The consequence of appending all earlier prompts and responses to each new prompt is that it consumes a lot of space in the context window. Therefore, some techniques have been developed to compress the information that is added to the user prompt. The simplest technique is to keep only the earlier user questions, but not the answers given by the application. In that case, for example, the modified prompt would be something like "*Earlier I said: 'How warm is Rio de Janeiro, Brazil, in the summer?', now answer only: "How long is the flight from New York to there?"*". Note that the prompt needs to tell the model to respond only to the last question submitted by the user.

Knowing that applications modify your prompts will be relevant for you if you test your prompts using consumer applications because the output you get from them can be substantially different from what you get when you use the model directly through an API, such as Microsoft Semantic Kernel. There's usually no way to know how the applications are modifying your prompts, as the providers usually don't reveal all their techniques.

Furthermore, a significant part of what you will do as an application developer will be to create prompt modifications that match your own application. Therefore, when your user submits their prompt, you will add your own prompt modifications to ensure they get an appropriate result. The techniques to modify user prompts are called **prompt engineering**, which we will explore briefly in the next chapter.

Generating text using consumer applications

Let's explore the two most popular text generation applications: ChatGPT and Microsoft Copilot. If you are not familiar with the power of GenAI, trying them out will give you a sense of what can be done with them and how powerful they are. We will also briefly talk about their configuration parameters and their architecture, which can help you decide the appropriate architecture for your own applications later.

OpenAI ChatGPT

If you have never used a GenAI app, ChatGPT is likely the best place to start. ChatGPT's default backend model is GPT 3.5, a fast and very powerful model. ChatGPT is free to use when OpenAI has available capacity on their servers. You can also buy a subscription to ChatGPT Plus for $20 per month, and that will give you the ability to use their most powerful model (currently GPT-4) and ensure you will always have capacity.

To use ChatGPT, go to `https://chat.openai.com`.

The ChatGPT interface is very simple. It allows you to choose the backend model on top, some suggestions of prompts in the middle, and a text box to enter prompts in the bottom. It also includes a notice that the output produced may include incorrect information:

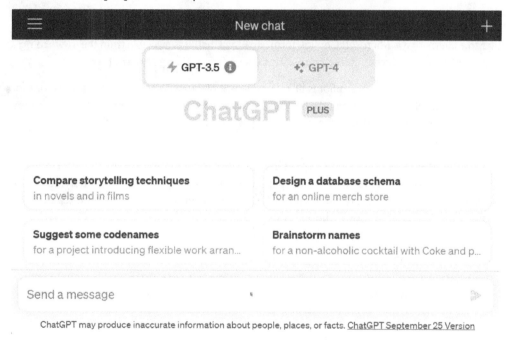

Figure 1.1 – Submitting requests to ChatGPT using the web interface

I will submit the following prompt in the **Send a Message** textbox:

```
How long is the flight between New York City and Rio de Janeiro?
```

ChatGPT, using the GPT-3.5 model, provides the following answer:

```
The flight duration between New York City and Rio de Janeiro can vary
depending on several factors, including the specific departure and
arrival airports, the flight route, and any layovers or stopovers.
On average, a non-stop flight between John F. Kennedy International
Airport (JFK) in New York City and Rio de Janeiro-Galeão International
Airport (GIG) in Rio de Janeiro takes approximately 9 to 10 hours.

Keep in mind that flight durations can vary, so it's always a good
idea to check with airlines for the most up-to-date and accurate
flight schedules and durations for your specific travel dates and
routes. Additionally, if you have layovers or stopovers, the total
travel time will be longer.
```

Microsoft Copilot

Another free alternative is Microsoft Copilot, formerly Bing Chat. It is available from the www.bing.com page, but it can be accessed directly from https://www.bing.com/chat.

The user interface for Microsoft Copilot is like the interface of ChatGPT. It has some suggestions for prompts in the middle of the screen and a text box, where the user can enter a prompt at the bottom. The Microsoft Copilot UI also shows a couple of options that will be relevant from when we use models programmatically.

The first is the conversation style. Copilot offers the option of being More Creative, More Balanced, or More Precise. This is related to the temperature parameter that will be passed to the underlying model. We will talk about the temperature parameter in *Chapter 3*, but in short, the temperature parameter determines how common the words chosen by the LLM are.

> ### Parameters in Microsoft Copilot
>
> While Microsoft Copilot does not reveal the exact configuration values, **Precise** is likely equivalent to a low temperature (maybe between 0 and 0.2), resulting in very safe guesses for the next word. For **Balanced**, the temperature is likely higher (maybe between 0.4 and 0.6), resulting *mostly* in safe guesses, but with the occasional guess being rare. **Creative** is the next step, likely around 0.8. Most guesses will still be safe, but more guesses will be rare words. Since LLMs guess words of a phrase in sequence, previous guesses influence subsequent guesses. When generating a phrase, each rare word makes the whole phrase more unusual.

Another interesting component in the UI is that the bottom right of the text box shows how many characters have been entered, giving you an idea of how much you will consume of the underlying model's context window. Note that you cannot know for sure how much you will consume because the Copilot application will modify your prompt.

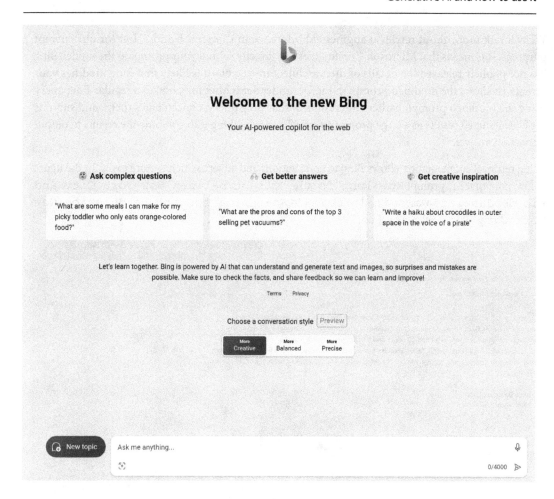

Figure 1.2 – Microsoft Copilot user interface

On August 11, 2023, Mikhail Parakhin, Bing's former CEO, posted on X/Twitter that Microsoft Copilot outperforms GPT-4 because it uses **retrieval augmented inference** (https://x.com/MParakhin/status/1689824478602424320?s=20):

Figure 1.3 – Post by Bing's former CEO about Microsoft Copilot using RAG

We will talk more about retrieval augmented inference in *Chapters 6* and *7*, but for our current purposes, this means that Microsoft Copilot does not directly submit your prompt to the model. Bing has not publicly released the details of their architecture yet, but it is likely that Bing modifies your prompt (it shows the modified prompt in the UI, under **Searching for**), makes a regular Bing query using the modified prompt, gathers the results of that Bing query, concatenates them, and submits the concatenated results as a large prompt to a GPT model, asking it to combine the results to output a coherent answer.

Using retrieval augmentation allows Bing to add citations and advertisements more easily. In the figure below, note that my prompt `How long is the flight between New York City and Rio de Janeiro?` was modified by Copilot to `Searching for flight duration New York City Rio de Janeiro`:

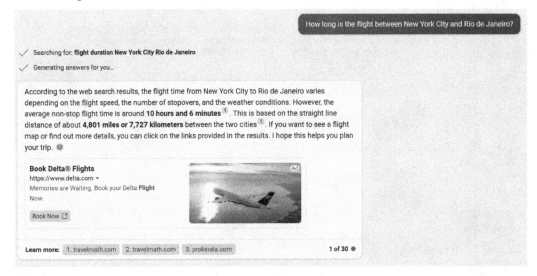

Figure 1.4 – An example of using Microsoft Copilot

As you can see, you can use consumer applications such as ChatGPT and Microsoft Copilot to familiarize yourself with how LLMs work for GenAI and to do some initial testing of your prompts, but be aware that the prompt you submit may be heavily modified by the application, and the response that you get from the underlying model can be very different from the responses that you will get when you actually create your own application.

Generating images

Besides generating text, AI can also be used to generate images from text prompts. The details of the process used to generate images from prompts are outside the scope of this book, but we will provide a quick summary. The main models in the image generation space are Midjourney, which is available through the Midjourney Bot in Discord; the open-source Stable Diffusion, which is also used by OpenAI's DALL-E 2; and the DALL-E 3, released in October 2023 and available through the Bing Chat (now Microsoft Copilot) and ChatGPT applications.

At the time of writing, Microsoft Semantic Kernel only supports DALL-E; therefore, this is the example we are going to explore. DALL-E 3 is available for free through the Microsoft Copilot application, with some limitations.

If you are using the Microsoft Copilot application from the earlier example, make sure to reset your chat history by clicking on the **New Topic** button to the left of the text box. To generate images, make sure your conversation style is set to **More Creative**, as image generation only works in Creative mode:

Figure 1.5 – Choosing the conversation style

I will use the following prompt:

```
create a photorealistic image of a salt-and-pepper standard schnauzer
on a street corner holding a sign "Will do tricks for cheese."
```

Microsoft Copilot will call DALL-E 3 and generate four images based on my specification:

Figure 1.6 – Images generated by Microsoft Copilot

One way in which DALL-E 3 is better than other image-generating models is that it can correctly add text to images. Earlier versions of DALL-E and most other models cannot spell words properly.

The images are presented in a grid with a total resolution of 1024 x 1024 pixels (512 x 512 per image). If you select one image, that specific image will be upscaled to the 1024 x 1024 pixel resolution. In my case, I will select the image in the bottom left corner. You can see the final result in the next figure:

Figure 1.7 – High-resolution image generated by Microsoft Copilot

As you can see, GenAI can also be used to generate images and now you have an idea of how powerful it can be. We will explore generating images with Microsoft Semantic Kernel in *Chapter 4*. There is a lot to explore before we get there, and we will start with a quick, comprehensive tour of Microsoft Semantic Kernel.

Microsoft Semantic Kernel

Microsoft Semantic Kernel (`https://github.com/microsoft/semantic-kernel`) is a thin, open source **software development toolkit** (**SDK**) that makes it easier for applications developed in C# and Python to interact with AI services such as the ones made available through OpenAI, Azure OpenAI, and Hugging Face. Semantic Kernel can receive requests from your application and route them to different AI services. Furthermore, if you extend the functionality of Semantic Kernel by adding your own functions, which we will explore in *Chapter 3*, Semantic Kernel can automatically discover which functions need to be used, and in which order, to fulfill a request. The request can come directly from the user and be passed through directly from your application, or your application can modify and enrich the user request before sending it to Semantic Kernel.

It was originally designed to power different versions of Microsoft Copilot, such as Microsoft 365 Copilot and the Bing Copilot, and then be released to the developer community as an open source package. Developers can use Semantic Kernel to create plugins that execute complex actions using AI services and combine these plugins using just a few lines of code.

In addition, Semantic Kernel can automatically orchestrate different plugins by using a **planner**. With the planner, a user can ask your application to achieve a complex goal. For example, if you have a function that identifies which animal is in a picture and another function that tells knock-knock jokes, your user can say, "*Tell me a knock-knock joke about the animal in the picture in this URL*," and the planner will automatically understand that it needs to call the identification function first and the "tell joke" function after it. Semantic Kernel will automatically search and combine your plugins to achieve that goal and create a plan. Then, Semantic Kernel will execute that plan and provide a response to the user:

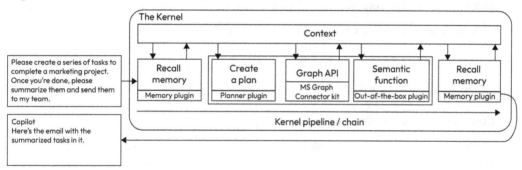

Figure 1.8 – Structure of Microsoft Semantic Kernel

In the upcoming sections, we will do a quick end-to-end tour of Semantic Kernel, going through most of the steps in *Figure 1.8*. We will send a request, create a plan, call the API, and call a native function and a semantic function. These, combined, will generate an answer to the user. First, we will do this manually, step-by-step, and then we will do everything in one go using the planner. You will see how powerful Semantic Kernel can be with just a little code.

Before we start experimenting with Microsoft Semantic Kernel, we need to install it.

Installing the Microsoft Semantic Kernel package

To use Microsoft Semantic Kernel, you must install it in your environment. Please note that Microsoft Semantic Kernel is still in active development, and there may be differences between versions.

Installing Microsoft Semantic Kernel in Python

To install Microsoft Semantic Kernel in Python, start in a new directory and follow these steps:

1. Create a new virtual environment with `venv`:

    ```
    python -m venv .venv
    ```

2. Activate the new environment you just created. This will ensure that Microsoft Semantic Kernel will be installed only for this directory:

    - In PowerShell, use the following:

        ```
        .venv/Scripts/activate.ps1
        ```

    - In Unix-like systems, such as Bash on Windows, MacOS, or Linux, use the following:

        ```
        .venv/Scripts/activate
        ```

3. Install Microsoft Semantic Kernel with `pip`:

    ```
    pip install semantic-kernel
    ```

Installing Microsoft Semantic Kernel in C#

To install Microsoft Semantic Kernel in C#, follow these steps:

1. Create a new project targeting .NET 8:

    ```
    dotnet new console -o ch1 -f net8.0
    ```

2. Change into the application directory:

    ```
    cd ch1
    ```

3. Add the `Microsoft.SemanticKernel` NuGet package:

    ```
    dotnet add package Microsoft.SemanticKernel --prerelease
    ```

    The kernel object itself is very lightweight. It is simply a repository of all the services and plugins that are connected to your application. Most applications start by instantiating an empty kernel and then adding services and functions to it.

Run the program with the following simple instructions just to make sure that the installation succeeded:

- Instantiating the kernel in Python:

    ```python
    import semantic_kernel as sk
    kernel = sk.Kernel()
    ```

- Instantiating the kernel in C#:

```
using Microsoft.SemanticKernel;

Kernel kernel = Kernel.CreateBuilder().Build()
```

We have now installed Semantic Kernel in your environment, and now we're ready to connect it to AI services and start using them.

Using Semantic Kernel to connect to AI services

To complete this section, you must have an API key. The process to obtain an API key was described at the beginning of this chapter.

In the upcoming subsections, we are only going to connect to the OpenAI text models GPT-3.5 and GPT-4. If you have access to the OpenAI models through Azure, you will need to make minor modifications to your code.

Although it would be simpler to connect to a single model, we are already going to show a simple but powerful Microsoft Semantic Kernel feature: we're going to connect to two different models and run a simple prompt using the simpler but less expensive model, GPT-3.5, and a more complex prompt on the more advanced but also more expensive model, GPT-4.

This process of sending simpler requests to simpler models and more complex requests to more complex models is something that you will frequently do when creating your own applications. This approach is called **LLM cascade**, and it was popularized in the FrugalGPT [3] paper. It can result in substantial cost savings.

> **Important: Order matters**
>
> The order in which you load your services matters. Both for Python (*step 3* in the *Connecting to OpenAI Services using Python* section) and for C# (*step 4* in the *Connecting to OpenAI Services using C#* section), we are going to first load the GPT-3.5 model into the kernel, followed by the GPT-4 model. This will make GPT-3.5 the default model. Later, we will specify which model will be used for which command; if we don't, GPT-3.5 will be used. If you load GPT-4 first, you will incur more costs.

We assume that you are using the OpenAI service instead of the Azure OpenAI service. You will need your OpenAI key and the organization ID, which can be found under **Settings** in the left menu of `https://platform.openai.com/account/api-keys`. All examples work with Azure OpenAI; you just need to use the Azure connection information instead of the OpenAI connection information.

Connecting to OpenAI Services using Python

This section assumes that you are using the OpenAI service. Before connecting to the OpenAI service, create a .env file in the ch1 directory containing your OpenAI key and your OpenAI organization ID. The organization ID can be found under Settings in the left menu of `https://platform.openai.com/account/api-keys`.

Your .env file should look like this, with the appropriate values replacing x in the following example:

```
OPENAI_API_KEY="sk-xxxxxxxxxxxxxxxxxxxxxxxxxxxxxxxxxxxxxxxx"
OPENAI_ORG_ID="org-xxxxxxxxxxxxxxxxxxxxxxx"
```

To connect to the OpenAI service using Python, perform the following steps:

1. Load an empty kernel:

    ```
    import semantic_kernel as sk
    kernel = sk.Kernel()
    ```

2. Load your API key and Organization ID into variables using the `openai_settings_from_dot_env` method from the `semantic_kernel_utils.settings` package:

    ```
    from semantic_kernel.utils.settings import openai_settings_from_
    dot_env
    api_key, org_id = openai_settings_from_dot_env()
    ```

3. Use the `OpenAIChatCompletion` method to create connections to chat services:

    ```
    from semantic_kernel.connectors.ai.open_ai import
    OpenAIChatCompletion
    gpt35 = OpenAIChatCompletion("gpt-3.5-turbo", api_key, org_id,
    service_id = "gpt35")
    gpt4 = OpenAIChatCompletion("gpt-4", api_key, org_id, service_id
    = "gpt4")
    kernel.add_service(gpt35)
    kernel.add_service(gpt4)
    ```

    If you are using OpenAI through Azure, instead of using `OpenAIChatCompletion`, you need to use `AzureOpenAIChatCompletion`, as shown here:

    ```
    kernel.add_service(
        AzureChatCompletion(
            service_id=service_id,
            deployment_name=deployment_name,
            endpoint=endpoint,
            api_key=api_key,
        ),
    )
    ```

Your Semantic Kernel is now ready to make calls, which we will do in the *Running a simple prompt* section.

Connecting to OpenAI services using C#

Before connecting to the OpenAI service, create a `config.json` file in the `ch1/config` directory, containing your OpenAI key and your OpenAI organization ID.

To avoid keeping a key in your code, we will load your keys from a configuration file. Your `config/settings.json` file should look like the following example, with the appropriate values in the `apiKey` and `orgId` fields (`orgId` is optional. If you don't have `orgId`, delete the field. An empty string will not work):

```
{
    "apiKey": "... your API key here ...",
    "orgId": "... your Organization ID here ..."
}
```

To connect to the OpenAI service in C#, perform the following steps:

1. Since we're going to reuse the API Key and Organization ID a lot, create a class to load them in `Settings.cs`:

```
using System.Text.Json;

public static class Settings {
  public static (string apiKey, string? orgId)
        LoadFromFile(string configFile = "config/settings.json")
    {
        if (!File.Exists(configFile))
        {
            Console.WriteLine("Configuration not found: " +
configFile);
            throw new Exception("Configuration not found");
        }
        try
        {
            var config = JsonSerializer.
Deserialize<Dictionary<string, string>>(File.
ReadAllText(configFile));

            // check whether config is null
            if (config == null)
            {
                Console.WriteLine("Configuration is null");
                throw new Exception("Configuration is null");
```

```
        }

        string apiKey = config["apiKey"];


        string? orgId;

        // check whether orgId is in the file
        if (!config.ContainsKey("orgId"))
        {
            orgId = null;
        }
        else
        {
            orgId = config["orgId"];
        }
        return (apiKey, orgId);
    }
    catch (Exception e)
    {
        Console.WriteLine("Something went wrong: " +
e.Message);
        return ("", "");
    }
  }
}
```

The preceding code is boilerplate code to read a JSON file and load its attributes into C# variables. We're looking for two attributes: apiKey and orgID.

2.   Load the settings from config/settings.json. We're going to create a class that will make this easier, as we are going to be doing this a lot. The class is very simple. It first checks whether the configuration file exists, and if it does, the class uses the JSON deserializer to load its contents into the apiKey and orgId variables:

```
using Microsoft.SemanticKernel;
var (apiKey, orgId) = Settings.LoadFromFile();
```

3.   Next, use the OpenAIChatCompletion method to create connections to chat services. Note that we're using serviceID to have a shortcut name for the model.

After you load the components, build the kernel with the Build method:

```
Kernel kernel = Kernel.CreateBuilder()
        .AddOpenAIChatCompletion("gpt-3.5-turbo", apiKey, orgId,
serviceId: "gpt3")
```

```
            .AddOpenAIChatCompletion("gpt-4", apiKey, orgId,
    serviceId: "gpt4")
                            .Build();
```

If you are using OpenAI through Azure, instead of using `AddOpenAIChatCompletion`, you need to use `AddAzureOpenAIChatCompletion`, as shown here:

```
Kernel kernel = Kernel.CreateBuilder()
                    .AddAzureOpenAIChatCompletion(modelId,
    endpoint, apiKey)
                    .Build();
```

Your Semantic Kernel is now ready to make calls, which we will do in the next section.

Running a simple prompt

This section assumes you completed the prior sections and builds upon the same code. By now, you should have instantiated Semantic Kernel and loaded both the GPT-3.5 and the GPT-4 services into it in that order. When you submit a prompt, it will default to the first service, and will run the prompt on GPT-3.5.

When we send the prompt to the service, we will also send a parameter called **temperature**. The temperature parameter goes from `0.0` to `1.0`, and it controls how random the responses are. We're going to explain the temperature parameter in more detail in later chapters. A temperature parameter of `0.8` generates a more creative response, while a temperature parameter of `0.2` generates a more precise response.

To send the prompt to the service, we will use a method called `create_semantic_function`. For now, don't worry about what a semantic function is. We're going to explain it in the *Using generative AI to solve simple problems* section.

Running a simple prompt in Python

To run a prompt in Python, follow these steps:

1. Load the prompt in a string variable:

    ```
    prompt = "Finish the following knock-knock joke. Knock, knock.
    Who's there? Dishes. Dishes who?"
    ```

2. Create a function by using the `add_function` method of the kernel. The `function_name` and `plugin_name` parameters are required but are not used, so you can give your function and plugin whatever name you want:

    ```
    prompt_function = kernel.add_function(function_name="ex01",
    plugin_name="sample", prompt=prompt)
    ```

3. Call the function. Note that all invocation methods are asynchronous, so you need to use `await` to wait for their return:

```
response = await kernel.invoke(prompt_function, request=prompt)
```

4. Print the response:

```
print(response)
```

The response is nondeterministic. Here's a possible response:

```
Dishes the police, open up!
```

Running a simple prompt in C#

To run a prompt in C#, follow these steps:

1. Load the prompt in a string variable:

```
string prompt = "Finish the following knock-knock joke. Knock,
knock. Who's there? Dishes. Dishes who?";
```

2. Call the prompt by using the `Kernel.InvokePromptAsync` function:

```
var joke = await kernel.InvokePromptAsync(prompt);
```

3. Print the response:

```
Console.Write(joke)
```

The response is nondeterministic. Here's a possible response:

```
Dishes a very bad joke, but I couldn't resist!
```

We have now connected to an AI service, submitted a prompt to it, and obtained a response. We're now ready to start creating our own functions.

Using generative AI to solve simple problems

Microsoft Semantic Kernel distinguishes between two types of functions that can be loaded into it: **semantic functions** and **native functions**.

Semantic functions are functions that connect to AI services, usually LLMs, to perform a task. The service is not part of your codebase. Native functions are regular functions written in the language of your application.

The reason to differentiate a native function from any other regular function in your code is that the native function will have additional attributes that will tell the kernel what it does. When you load a native function into the kernel, you can use it in chains that combine native and semantic functions. In addition, Semantic Kernel planner can use the function when creating a plan to achieve a user goal.

Creating semantic functions

We have already created a semantic function (knock) in the previous section. Now, we're going to add a parameter to it. The default parameter for all semantic functions is called {{$input}}.

Modified semantic function in Python

We're going to make minor modifications to our previous code to allow the semantic function to receive a parameter. Again, the following code assumes that you have already instantiated a kernel and connected to at least one service:

```
from semantic_kernel.functions.kernel_arguments import
KernelArguments
args = KernelArguments(input="Boo")
response = await kernel.invoke(prompt_function, request=prompt,
arguments=args)
print(response)
```

The only differences from the code before are that now we have a variable, {{$input}}, and we're calling the function using a parameter, the string "Boo". To add the variable, we need to import the KernelArguments class from the semantic_kernel_functions.kernel_arguments package and create an instance of the object with the value we want.

The answer is nondeterministic. Here's a possible answer:

```
Don't cry, it's just a joke!
```

Modified semantic function in C#

To create a function in C#, we are going to use the CreateFunctionFromPrompt kernel method, and to add a parameter, we will use the KernelArguments object:

```
string prompt = "Finish the following knock-knock joke. Knock, knock.
Who's there? {{$input}}, {{$input}} who?";
KernelFunction jokeFunction = kernel.CreateFunctionFromPrompt(prompt);
var arguments = new KernelArguments() { ["input"] = "Boo" };

var joke = await kernel.InvokeAsync(jokeFunction, arguments);

Console.WriteLine(joke);
```

Here, too, the only differences from the code before are that now we have a variable, `{{$input}}`, and we're calling the function using a parameter, the string `"Boo"`.

The answer is nondeterministic. Here's a possible answer:

```
Don't cry, it's just a joke!
```

Creating native functions

Native functions are created in the same language your application is using. For example, if you are writing code in Python, a native function can be written in Python.

Although you can call a native function directly without loading it into the kernel, loading makes it available to the planner, which we will see in the last section of this chapter.

We're going to explore native functions in greater detail in *Chapter 3*, but for now, let's create and load a simple native function in the Kernel.

The native function we're going to create chooses a theme for a joke. For now, the themes are `Boo`, `Dishes`, `Art`, `Needle`, `Tank`, and `Police`, and the function simply returns one of these themes at random.

Creating a native function in Python

In Python, the native functions need to be inside a class. The class used to be called a **skill**, and in some places, this name is still used. The name has recently changed to **plugin**. A plugin (formerly called skill) is just a collection of functions. You cannot mix native and semantic functions in the same skill.

We're going to name our class `ShowManager`.

To create a native function, you will use the `@kernel_function` decorator. The decorator must contain fields for `description` and `name`. To add a decorator, you must import `kernel_function` from the `semantic_kernel.functions.kernel_function_decorator` package.

The function body comes immediately after the decorator. In our case, we are simply going to have the list of themes and use the `random.choice` function to return one random element from the list:

```python
import random
class ShowManager():
    @kernel_function(
    description="Randomly choose among a theme for a joke",
    name="random_theme"
    )
    def random_theme(self) -> str:
        themes = ["Boo", "Dishes", "Art",
                "Needle", "Tank", "Police"]
        theme = random.choice(themes)
        return theme
```

Then, to load the plugin and all its functions in the kernel, we use the add_plugin method of the kernel. When you are adding a plugin, you need to give it a name:

```
theme_choice = kernel.add_plugin(ShowManager(), "ShowManager")
```

To call the native function from a plugin, simply put the name of the function within brackets, as shown here:

```
response = await kernel.invoke(theme_choice["random_theme"])
print(response)
```

The function is not deterministic, but a possible result might be:

```
Tank
```

Creating a native function in C#

In C#, native functions need to be inside a class. The class used to be called a skill, and this name is still used in some places; for example, in the SDK, we will need to import Microsoft.SemanticKernel. SkillDefinition. Skills have recently been renamed to plugins. A plugin is just a collection of functions. You cannot mix native and semantic functions in the same skill.

We're going to name our class ShowManager.

To create a native function, you will use the [KernelFunction] decorator. The decorator must contain Description. The function body comes immediately after the decorator. In our case, we are simply going to have a list of themes and use the Random().Next method to return one random element from the list. We will call our class ShowManager and our function RandomTheme:

```
using System.ComponentModel;
using Microsoft.SemanticKernel;

namespace Plugins;

public class ShowManager
{
    [KernelFunction, Description("Take the square root of a number")]
    public string RandomTheme()
    {
        var list = new List<string> { "boo", "dishes", "art",
"needle", "tank", "police"};
        return list[new Random().Next(0, list.Count)];
    }
}
```

Then, to load the plugin and all its functions into the kernel, we use the
`ImportPluginFromObject` method:

```
string prompt = "Finish the following knock-knock joke. Knock, knock.
Who's there? {{$input}}, {{$input}} who?";
KernelFunction jokeFunction = kernel.CreateFunctionFromPrompt(prompt);
var showManagerPlugin = kernel.ImportPluginFromObject(new Plugins.
ShowManager());
var joke = await kernel.InvokeAsync(jokeFunction, arguments);
Console.WriteLine(joke);
```

To call the native function from a plugin, simply put the function name within brackets. You can pass parameters by using the `KernelArguments` class, as shown here:

```
var result = await kernel.
InvokeAsync(showManagerPlugin["RandomTheme"]);
Console.WriteLine("I will tell a joke about " + result);

var arguments = new KernelArguments() { ["input"] = result };
```

The function is not deterministic, but a possible result might be the following:

```
I will tell a joke about art
```

Now that you can run simple prompts from your code, let's learn to separate the prompt configuration from the code that calls it by using plugins.

Plugins

One of the greatest strengths of Microsoft Semantic Kernel is that you can create semantic plugins that are language agnostic. Semantic plugins are collections of semantic functions that can be imported into the kernel. Creating semantic plugins allows you to separate your code from the AI function, which makes your application easier to maintain. It also allows other people to work on the prompts, making it easier to implement prompt engineering, which will be explored in *Chapter 2*.

Each function is defined by a directory containing two text files: `config.json`, which contains the configuration for the semantic function, and `skprompt.txt`, which contains its prompt.

The configuration of the semantic function includes the preferred engine to use, the temperature parameter, and a description of what the semantic function does and its inputs.

The text file contains the prompt that will be sent to the AI service to generate the response.

In this section, we are going to define a plugin that contains two semantic functions. The first semantic function is a familiar function: the knock-knock joke generator. The second function is a function that receives a joke as an input and tries to explain why it's funny. Since this is a more complicated task, we're going to use GPT-4 for this.

```
Let's take a look at the directory structure: └──plugins
    └──jokes
        |──knock_knock_joke
        |    ├──config.json
        |    └──skprompt.txt
        ├──explain_joke
            ├──config.json
            └──skprompt.txt
```

We will now see how to create the `config.json` and `skprompt.txt` files and how to load the plugin into our program.

The config.json file for the knock-knock joke function

The following configuration file shows a possible configuration for the semantic function that generates knock-knock jokes:

```
{
    "schema": 1,
    "type": "completion",
    "description": "Generates a knock-knock joke based on user input",
    "default_services": [
        "gpt35",
        "gpt4"
    ],
    "execution_settings": {
        "default": {
            "temperature": 0.8,
            "number_of_responses": 1,
            "top_p": 1,
            "max_tokens": 4000,
            "presence_penalty": 0.0,
            "frequency_penalty": 0.0
        }
    },
    "input_variables": [
```

```
        {
            "name": "input",
            "description": "The topic that the joke should be written
about",
            "required": true
        }
    ]
}
```

The `default_services` property is an array of the preferred engines to use (in order). Since knock-knock jokes are simple, we're going to use GPT-3.5 for it. All the parameters in the preceding file are required. In future chapters, we will explain each parameter in detail, but for now, you should just copy them.

The `description` field is important because it can be used later by the planner, which will be explained in the last section of this chapter.

The skprompt.txt file for the knock-knock joke function

Since we want to explain the joke later, we need our application to return the whole joke, not only the punchline. This will enable us to save the whole joke and pass it as a parameter to the explain-the-joke function. To do so, we need to modify the prompt. You can see the final prompt here:

```
You are given a joke with the following setup:

Knock, knock!
Who's there?
{{$input}}!
{{$input}} who?
Repeat the whole setup and finish the joke with a funny punchline.
```

The config.json file for the semantic function that explains jokes

You should now create a file for the function that explains jokes. Since this is a more complicated task, we should set `default_services` to use GPT-4.

This file is almost exactly the same as the `config.json` file used for the knock-knock joke function. We have made only three changes:

- The description
- The description of the `input` variable
- The `default_services` field

This can be seen in the following:

```
{
    "schema": 1,
    "type": "completion",
    "description": "Given a joke, explain why it is funny",
    "default_services": [
        "gpt4"
    ],
    "execution_settings": {
        "default": {
            "temperature": 0.8,
            "number_of_responses": 1,
            "top_p": 1,
            "max_tokens": 4000,
            "presence_penalty": 0.0,
            "frequency_penalty": 0.0
        }
    },
    "input_variables": [
        {
            "name": "input",
            "description": "The joke that we want explained",
            "required": true
        }
    ]
}
```

The skprompt.txt file for the explain joke function

The prompt for the function that explains jokes is very simple:

```
You are given the following joke:
{{$input}}
First, tell the joke.
Then, explain the joke.
```

Loading the plugin from a directory into the kernel

Now that the semantic functions are defined in text files, you can load them into the kernel by simply pointing to the directory where they are. This can also help you to separate the prompt engineering function from the development function. Prompt engineers can work with the text files without ever having to touch the code of your application.

Loading the plugin using Python

You can load all the functions inside a plugin directory using the `add_plugin` method from the kernel object. Just set the first parameter to `None` and set the `parent_directory` parameter to the directory where the plugin is:

```
jokes_plugin = kernel.add_plugin(None, parent_directory="../../
plugins", plugin_name="jokes")
```

You can call the functions in the same way as you would call a function from a native plugin by putting the function name within brackets:

```
knock_joke = await kernel.invoke(jokes_plugin["knock_knock_joke"],
KernelArguments(input=theme))
print(knock_joke)
```

The result of the preceding call is nondeterministic. Here's a sample result:

```
Knock, knock!
Who's there?
Dishes!
Dishes who?
Dishes the police, open up, we've got some dirty plates to wash!
```

We can pass the results of the preceding call to the `explain_joke` function:

```
explanation = await kernel.invoke(jokes_plugin["explain_joke"],
KernelArguments(input=knock_joke))
print(explanation)
```

Remember that this function is configured to use GPT-4. The results of this function are nondeterministic. Here's a sample result:

```
This joke is funny because it plays off the expectation set by
the traditional "knock, knock" joke format. Typically, the person
responding sets up a pun or a simple joke with their question ("...
who?"), but instead, the punchline in this joke is a whimsical and
unexpected twist: the police are here not to arrest someone, but
to wash dirty plates. This absurdity creates humor. Also, the word
'dishes' is used in a punning manner to sound like 'this is'.
```

Loading the plugin using C#

You can load all the functions inside a plugin directory. First, we obtain the path to the directory (your path may be different):

```
var pluginsDirectory = Path.Combine(System.IO.Directory.
GetCurrentDirectory(),
        "..", "..", "..", "plugins", "jokes");
```

Then, we use `ImportPluginFromPromptDirectory` to load the functions into a variable. The result is a collection of functions. You can access them by referencing them inside brackets:

```
var jokesPlugin = kernel.
ImportPluginFromPromptDirectory(pluginsDirectory, "jokes");
```

The last step is to call the function. To call it, we use the `InvokeAsync` method of the kernel object. We will, again, pass a parameter using the `KernelArguments` class:

```
var result = await kernel.InvokeAsync(jokesPlugin["knock_knock_joke"],
new KernelArguments() {["input"] = theme.ToString()});)
```

The result of the preceding call is nondeterministic. Here's a sample result:

```
Knock, knock!
Who's there?
Dishes!
Dishes who?
Dishes the best joke you've heard in a while!
```

To get an explanation, we can pass the results of the preceding call to the `explain_joke` function:

```
var explanation = await kernel.InvokeAsync(jokesPlugin["explain_
joke"], new KernelArguments() {["input"] = result});
Console.WriteLine(explanation);
```

Here's a sample result:

```
Knock, knock!
Who's there?
Dishes!
Dishes who?
Dishes the best joke you've heard in a while!
Now, let's break down the joke:
The joke is a play on words and relies on a pun. The setup follows
the classic knock, knock joke format, with the person telling the joke
pretending to be at the door. In this case, they say "Dishes" when
asked who's there.
Now, the pun comes into play when the second person asks "Dishes who?"
Here, the word "Dishes" sounds similar to the phrase "This is." So, it
can be interpreted as the person saying "This is the best joke you've
heard in a while!"
The punchline subverts the expectation of a traditional knock, knock
joke response, leading to a humorous twist. It plays on the double
meaning of the word "Dishes" and brings humor through wordplay and
cleverness.
```

Now that you have seen how to create and call one function of a plugin, we are going to learn how to use a planner to call multiple functions from different plugins.

Using a planner to run a multistep task

Instead of calling functions yourself, you can let Microsoft Semantic Kernel choose the functions for you. This can make your code a lot simpler and can give your users the ability to combine your code in ways that you haven't considered.

Right now, this will not seem very useful because we only have a few functions and plugins. However, in a large application, such as Microsoft Office, you may have hundreds or even thousands of plugins, and your users may want to combine them in ways that you can't yet imagine. For example, you may be creating a copilot that helps a user be more efficient when learning about a subject, so you write a function that downloads the latest news about that subject from the web. You may also have independently created a function that explains a piece of text to the user so that the user can paste content to learn more about it. The user may decide to combine them both with "*download the news and write an article explaining them to me*," something that you never thought about and didn't add to your code. Semantic Kernel will understand that it can call the two functions you wrote in sequence to complete that task.

When you let users request their own tasks, they will use natural language, and you can let Semantic Kernel inspect all the functions that are loaded into it and use a planner to decide the best way of handling the user request.

For now, we are only going to show a quick example of using a planner, but we will explore the topic in more depth in *Chapter 5*. Planners are still under active development, and there might be changes over time. Currently, Semantic Kernel is expected to have two planners: the **Function Calling Stepwise planner**, available for Python and C#, and the **Handlebars planner**, available only for C# at the time of writing.

Although the following example is very simple and both planners behave in the same way, we will show how to use the Stepwise planner (Function Calling Stepwise Planner) with Python and the Handlebars planner with C#.

Calling the Function Calling Stepwise planner with Python

To use the Stepwise planner, we first create an object of the `FunctionCallingStepwisePlanner` class and make a request to it. In our case, we're going to ask it to choose a random theme, create a knock-knock joke, and explain it.

We're going to modify our earlier program, delete the function calls, and add a call to the planner instead:

```
ask = f"""Choose a random theme for a joke, generate a knock-knock
joke about it and explain it"""
options = FunctionCallingStepwisePlannerOptions(
```

```
   max_iterations=10,
   max_tokens=4000)
planner = FunctionCallingStepwisePlanner(service_id="gpt4",
options=options)

result = await planner.invoke(kernel, ask)
print(result.final_answer)
```

There are a couple of details to note. The first one is that I used the class `FunctionCallingStepwisePlannerOptions` to pass a `max_tokens` parameter to the planner. Behind the scenes, the planner will create a prompt and send it to the AI service. The default `max_tokens` for most AI services tends to be small. At the time of writing, it was `250`, which may cause an error if the prompt generated by the planner is too large. The second detail to note is that I printed `result.final_answer` instead of `result`. The `result` variable contains the whole plan: the definition of the functions, the chat with the OpenAI model explaining how to proceed, etc. It's interesting to print the `result` variable to see how the planner works internally, but to see the outcome of the planner execution, all you need to do is print `result.final_answer`.

Here is a sample response, first telling the joke and then explaining it:

```
First, the joke:
Knock, knock!
Who's there?
Police!
Police let me in, it's cold out here!
Now, the explanation:
The humor in this joke comes from the play on words. The word "police"
is being used in a different context than typically used. Instead
of referring to law enforcement, it's used as a pun to sound like
"Please". So, when the jokester says "Police let me in, it's cold out
here!", it sounds like "Please let me in, it's cold out here!". Knock,
knock jokes are a form of humor that relies on word play and puns, and
this joke is a standard example of that.
```

As you can see, the planner generated the joke and the explanation, as expected, without us needing to tell Semantic Kernel in which order to call the functions.

Calling the Handlebars planner in C#

At the time of writing, the Handlebars planner is in version 1.0.1-preview, and it's still experimental in C#, although it's likely that a release version will be made available soon.

To use the Handlebars planner, you first need to install it, which you can do by using the following command (you should use the latest version available to you):

```
dotnet add package Microsoft.SemanticKernel.
s.Handlebars --version 1.0.1-preview
```

To use the Handlebars planner, you need to use the following `pragma` warning in your code. The Handlebars planner code is still experimental, and if you don't add the `#pragma` directive, your code will fail, with a warning that it contains experimental code. You also need to import the `Microsoft. SemanticKernel.Planning.Handlebars` package:

```
#pragma warning disable SKEXP0060
using Microsoft.SemanticKernel;
using Microsoft.SemanticKernel.Planning.Handlebars;
```

We proceed as usual, instantiating our kernel and adding native and semantic functions to it:

```
var (apiKey, orgId) = Settings.LoadFromFile();
Kernel kernel = Kernel.CreateBuilder()
        .AddOpenAIChatCompletion("gpt-3.5-turbo", apiKey, orgId,
serviceId: "gpt3")
        .AddOpenAIChatCompletion("gpt-4", apiKey, orgId, serviceId:
"gpt4")
                        .Build();
var pluginsDirectory = Path.Combine(System.IO.Directory.
GetCurrentDirectory(),
        "..", "..", "..", "plugins", "jokes");
```

The big difference happens now – instead of telling which functions to call and how, we simply ask the planner to do what we want:

```
var goalFromUser = "Choose a random theme for a joke, generate a
knock-knock joke about it and explain it";
var planner = new HandlebarsPlanner
(new HandlebarsPlannerOptions() { AllowLoops = false });
var plan = await
planner.CreatePlanAsync(kernel, goalFromUser);
```

We can execute the plan by calling `InvokeAsync` from the `plan` object:

```
var result = await plan.InvokeAsync(kernel);
Console.WriteLine(result);
```

The result is nondeterministic. Here is a sample result, first telling the joke and then explaining it:

```
Knock, knock!
Who's there?
Police!
Police who?
Police let me know if you find my sense of humor arresting!

Explanation:
```

```
This joke is a play on words and relies on the double meaning of the
word "police."

In the setup, the person telling the joke says "Knock, knock!" which
is a common way to begin a joke. The other person asks "Who's there?"
which is the expected response.

The person telling the joke then says "Police!" as the punchline,
which is a word that sounds like "please." So it seems as if they are
saying "Please who?" instead of "Police who?"

Finally, the person telling the joke completes the punchline by saying
"Police let me know if you find my sense of humor arresting!" This is
a play on words because "arresting" can mean two things: first, it can
mean being taken into custody by the police, and second, it can mean
captivating or funny. So the person is asking if the listener finds
their sense of humor funny or engaging and is also using the word
"police" to continue the play on words.
```

As you can see, the planner generated the joke and the explanation, as expected, without us needing to tell Semantic Kernel in which order to call the functions.

Summary

In this chapter, you learned about Generative AI and the main components of Microsoft Semantic Kernel. You learned how to create a prompt and submit it to a service and how to embed that prompt into a semantic function. You also learned how to execute multistep requests by using a planner.

In the next chapter, we are going to learn how to make our prompts better through a topic called **prompt engineering**. This will help you create prompts that get your users the correct result faster and use fewer tokens, therefore reducing costs.

References

[1] A. Vaswani et al., "Attention Is All You Need," Jun. 2017.

[2] OpenAI, "GPT-4 Technical Report." arXiv, Mar. 27, 2023. doi: 10.48550/arXiv.2303.08774.

[3] L. Chen, M. Zaharia, and J. Zou, "FrugalGPT: How to Use Large Language Models While Reducing Cost and Improving Performance." arXiv, May 09, 2023. doi: 10.48550/arXiv.2305.05176.

2
Creating Better Prompts

As a developer, you can request that an LLM completes a task by submitting a prompt to it. In the previous chapter, we saw some examples of prompts, such as "*Tell me a knock-knock joke*" and "*What is the flight duration between New York City and Rio de Janeiro?*" As LLMs became more powerful, the tasks that they could accomplish became more complex.

Researchers discovered that using different techniques to build prompts yielded vastly different results. The process of crafting prompts that improve the likelihood of getting the desired answer is called prompt engineering, and the value of creating better prompts gave birth to a new profession: **prompt engineer**. This is someone who doesn't need to know how to code in any programming language but can create prompts using natural language that return the desired results.

Microsoft Semantic Kernel uses the concept of **prompt templating**, the creation of structured templates for prompts that contain placeholders for specific types of information and instructions that can be filled in or customized by the user or by the developer. By using prompt templates, developers can introduce multiple variables in prompts, separate the prompt engineering function from the coding function, and use advanced prompt techniques to increase accuracy in responses.

In this chapter, you'll learn about several techniques that will make your prompts more likely to return the results you want your users to see in the first attempt. You'll learn how to employ prompts with multiple variables as well as how to create and use prompts that have multiple parameters to complete more complex tasks. Finally, you'll uncover techniques that combine prompts in creative ways to improve accuracy in scenarios where LLMs are not very accurate – for example, when solving math problems.

In this chapter, we'll be covering the following topics:

- Engineering prompts
- Prompts with multiple variables
- Prompts with multiple stages

Technical requirements

To complete this chapter, you will need to have a recent, supported version of your preferred Python or C# development environment:

- For Python, the minimum supported version is Python 3.10, and the recommended version is Python 3.11
- For C#, the minimum supported version is .NET 8

You will also need an **OpenAI API** key, obtained either directly through **OpenAI** or through **Microsoft**, through the **Azure OpenAI** service. Instructions on how to obtain these keys can be found in *Chapter 1*.

If you are using .NET, the code for this chapter can be found at `https://github.com/PacktPublishing/Building-AI-Applications-with-Microsoft-Semantic-Kernel/tree/main/dotnet/ch2`.

If you are using Python, the code for this chapter can be found at `https://github.com/PacktPublishing/Building-AI-Applications-with-Microsoft-Semantic-Kernel/tree/main/python/ch2`.

You can install the required packages by going to the GitHub repository and using the following: `pip install -r requirements.txt`.

A simple plugin template

There are two simple ways of creating prompt templates.

The first is to generate the prompt from a string variable inside your code. This way is simple and convenient. We covered this method in the *Running a simple prompt* section of *Chapter 1*.

The second is to use Semantic Kernel to help separate the development function from the prompt engineering function. As you saw in *Chapter 1*, you can create your requests to LLMs as functions in plugins. A plugin is a directory that contains multiple sub-directories, one per function. Each subdirectory will have exactly two files:

- A text file called `skprompt.txt` that contains the prompt
- A configuration file called `config.json` that contains the parameters that will be used in the API call

Since the prompt is maintained separately from the code, you, as an application developer, can focus on the code of your application and let a specialized prompt engineer work on the `skprompt.txt` files.

In this chapter, we will focus on the second method – creating plugins in dedicated directories – because this method is more robust to changes. For example, if you are switching from Python to C# or using a new version of the .NET library and these changes require a lot of changes to be made to your code, at least you don't need to change your prompts and function configurations.

The code for this prompt plugin is like the one we used in the previous chapter. We are going to take the plugin we built in *Chapter 1* and make several changes to `skprompt.txt` to observe the results and learn how different prompting techniques can substantially change outcomes.

We are doing this again in detail at the beginning of this chapter even though we did something very similar in *Chapter 1* because we're going to use these steps several times as we explore prompt engineering.

For the examples we'll cover here, you can use both GPT-3.5 and GPT-4, but remember that GPT-4 is 30x more expensive. The results that are shown are from GPT-3.5 unless indicated otherwise.

The skprompt.txt file

We will start with a very simple prompt, just asking directly what we want, without any additional context:

```
Create an itinerary of three must-see attractions in {{$city}}.
```

That's it – that's the whole file.

The config.json file

The `config.json` file is very similar to the one we used in *Chapter 1*, but we changed three things:

- The description of the function
- The name of the input variable under `input_variables`
- The description of the input variable under `input_variables`

```
{
    "schema": 1,
    "type": "completion",
    "description": "Creates a list of three must-see attractions for
someone traveling to a city",
    "default_services": [
        "gpt35"
    ],
    "execution_settings": {
        "default": {
            "temperature": 0.8,
            "number_of_responses": 1,
            "top_p": 1,
            "max_tokens": 4000,
            "presence_penalty": 0.0,
            "frequency_penalty": 0.0
```

```
                }
        },
        "input_variables": [
            {
                "name": "city",
                "description": "The city the person wants to travel to",
                "required": true
            }
        ]
}
```

Now that we have defined the function, let's call it.

Calling the plugin from Python

The code we're using here is very similar to what we used in *Chapter 1*. We've made three small changes: use only GPT-3.5, point to the appropriate plugin directory, `prompt_engineering`, and change the input variable's name from `input` to `city`:

```python
import asyncio
from semantic_kernel.connectors.ai.open_ai import OpenAIChatCompletion
from semantic_kernel.utils.settings import openai_settings_from_dot_
env
import semantic_kernel as sk
from semantic_kernel.functions.kernel_arguments import KernelArguments

async def main():
    kernel = sk.Kernel()
    api_key, org_id = openai_settings_from_dot_env()
    gpt35 = OpenAIChatCompletion("gpt-3.5-turbo", api_key, org_id,
"gpt35")

    kernel.add_service(gpt35)

    pe_plugin = kernel.add_plugin(None, parent_directory="../../
plugins", plugin_name="prompt_engineering")
    response = await kernel.invoke(pe_plugin["attractions_single_
variable"], KernelArguments(city="New York City"))
    print(response)


if __name__ == "__main__":
    asyncio.run(main())
```

Now, let's learn how to call the plugin in C#. Note that we don't make any changes to `skprompt.txt` or `config.json`. We can use the same prompt and configuration, independent of the language.

Calling the plugin from C#

As we did in Python, we just need to make three small changes: use only GPT-3.5, point to the appropriate plugin directory, `prompt_engineering`, and change the input variable's name from `input` to `city`:

```
using Microsoft.SemanticKernel;

var (apiKey, orgId) = Settings.LoadFromFile();

Kernel kernel = Kernel.CreateBuilder()
                        .AddOpenAIChatCompletion("gpt-3.5-turbo",
apiKey, orgId, serviceId: "gpt35")
                        .Build();


var pluginsDirectory = Path.Combine(System.IO.Directory.
GetCurrentDirectory(),
        "..", "..", "..", "plugins", "prompt_engineering");

var promptPlugin = kernel.
ImportPluginFromPromptDirectory(pluginsDirectory, "prompt_
engineering");
var result = await kernel.InvokeAsync(promptPlugin["attractions_
single_variable"], new KernelArguments() {["city"] = "New York
City"});

Console.WriteLine(result);
```

Let's see the results.

Results

Instead of giving the top three attractions that I wanted to see, in one of the calls, GPT-3.5 created a 3-day list of attractions and gave me a list of six attractions. This is a typical problem that can happen when the prompt is not very specific. The list of attractions is very good and considers how hard it is to move around in New York City due to traffic: attractions on the same day are close to each other, or at least there is enough time to travel between them. Since a lot of GPT-3.5 training was done with data during the COVID period, the itinerary even includes a note to check for COVID closures.

Note that your result may be different, and in some cases, you may get a response that looks like the original intent, with only three must-see attractions:

```
Itinerary: Must-See Attractions in New York City


Day 1:
1. Statue of Liberty and Ellis Island: Start your visit to New York
City by taking a ferry to the Statue of Liberty and Ellis Island.
Explore the grounds of Lady Liberty, marvel at this iconic symbol
of freedom and take in stunning views of the city skyline from the
observation deck. Then, head to Ellis Island to learn about the
millions of immigrants who passed through its gates and contributed to
the cultural fabric of America.


2. Times Square: In the afternoon, immerse yourself in the vibrant
energy of Times Square. This bustling, neon-lit intersection is known
for its towering billboards, dazzling Broadway theaters, and bustling
crowds. Take a stroll along the pedestrian-friendly plazas, snap
photos in front of iconic landmarks like the TKTS booth, and soak up
the excitement of this iconic New York City landmark.


Day 2:
3. The Metropolitan Museum of Art (The Met): Spend the morning
exploring one of the world's largest and most famous art museums, The
Met. This cultural treasure houses an extensive collection spanning
5,000 years of human history, encompassing art from various regions
and civilizations. Admire masterpieces from ancient Egypt, classical
antiquity, European Renaissance artists, and contemporary art. Don't
miss the rooftop garden for panoramic views of Central Park and the
Manhattan skyline.


4. Central Park: After the museum, take a leisurely stroll through
Central Park, an urban oasis in the heart of the city. This sprawling
green space offers a refreshing break from the bustling streets. Enjoy
a picnic near the Bethesda Terrace and Fountain, rent a rowboat on the
lake, visit the Central Park Zoo, or simply relax and people-watch in
this iconic park.


Day 3:
5. The High Line: Start your day with a visit to the High Line, a
unique elevated park built on a historic freight rail line. This
linear park stretches for 1.45 miles and offers stunning views of the
city, beautifully landscaped gardens, public art installations, and a
variety of seating areas. Take a leisurely walk along the promenade,
enjoy the greenery, and appreciate the innovative urban design.


6. The 9/11 Memorial & Museum: In the afternoon, visit the 9/11
Memorial & Museum, a deeply moving tribute to the victims of the
September 11, 2001, terrorist attacks. The memorial features two
reflecting pools set in the footprints of the Twin Towers, surrounded
by bronze parapets inscribed with the names of those who lost their
```

```
lives. Inside the museum, you'll find artifacts, multimedia displays,
and personal accounts that document the events and aftermath of 9/11.

Note: This itinerary includes popular attractions in New York City,
but it's always worth checking for any COVID-19-related restrictions
or closures before planning your visit.
```

This quick, simple prompt didn't work consistently as expected, so let's explore how to make things better by adding more information to the prompt.

Improving the prompt to get better results

If the main way you interact with LLMs is using ChatGPT or Microsoft Copilot, you may have the impression that you can use very short prompts to get the results you want. As explained in *Chapter 1*, these applications modify your submission and add a lot of instructions to your prompt. As a developer, you'll have to do the same. Here are a few tips to improve your prompt and obtain better results:

- **Provide context**: Instead of simply asking for three must-see attractions, provide as much context as you can. Think about everything that you would ask someone who asked you the same question – for example, "*How many days are you staying?*," "*What kind of things do you like and dislike?*," or "*How are you going to be getting around?*"

- **Be explicit about the format of the response**: If you want the answer to come out in a specific format, make sure you tell the LLM this. You may even provide some examples – for example, "*Answer with a single word – 'Yes' or 'No.'*"

- **Specify the response's length**: If you want a concise answer, specify the desired length – for example, "*In two sentences.*" If you want a longer answer, make it clear – for example, "*600 words*" or "*five paragraphs.*"

Revising the skprompt.txt file

Using these hints, we're going to rewrite the prompt to see if we can get improved responses consistently:

```
I am traveling to {{$city}} and staying only 24 hours.
I am afraid of heights.
I will be downtown.
I like movies and TV shows a lot.
I am traveling by myself.
I will get around using public transportation or Uber.
I will be there on a Saturday.
Create an itinerary of three attractions for me to visit.
The attractions should be listed in the following format:
- Attraction name: [name]
- What is the best thing to see there: [one sentence description]
Your whole answer should be less than 500 words.
```

The result

Here's the response we get from the prompt:

```
Itinerary for a 24-hour trip to New York City:
Attraction 1: Times Square
- What is the best thing to see there: The vibrant neon lights,
billboards, and bustling atmosphere of one of the world's most famous
intersections.
Start your day by visiting Times Square, the iconic heart of New York
City. Immerse yourself in the vibrant energy and larger-than-life
atmosphere of this bustling area. Marvel at the enormous billboards
and screens displaying advertisements for Broadway shows, movies,
and various brands. While you're there, make sure to take a stroll
along the pedestrian plazas, people-watch, and soak in the bustling
activity. Don't forget to snap some photos to capture the essence of
this iconic location.
Attraction 2: Central Park
- What is the best thing to see there: Explore the beautiful
green landscape, visit popular landmarks like Bethesda Terrace and
Strawberry Fields, and enjoy a peaceful escape from the city.
Escape the concrete jungle and head to Central Park, an oasis of
tranquility and natural beauty right in the heart of Manhattan.
Take a leisurely walk through the park's lush greenery, visit famous
spots such as Bethesda Terrace and Bethesda Fountain, and enjoy the
peacefulness that surrounds you. If you're a fan of movies or TV
shows, keep an eye out for locations you may recognize from various
film shoots. Grab a coffee or snack from one of the park's vendors and
find a cozy spot to relax and people-watch, or rent a bike and explore
the park's vast expanse at your own pace.
Attraction 3: Museum of the Moving Image
- What is the best thing to see there: Delve into the world of movies
and TV shows through interactive exhibits, iconic props, and behind-
the-scenes insights.

As a movie and TV show enthusiast, a visit to the Museum of the Moving
Image in Queens is a must. This unique museum offers a fascinating
exploration of the history, art, and technology behind the moving
image. Get a glimpse of iconic props, costumes, and memorabilia
from beloved films and TV shows. Engage in interactive exhibits that
allow you to create your own animations or dub your voice into famous
movie scenes. Learn about the evolution of filmmaking techniques and
gain behind-the-scenes insights into the process of creating the
entertainment you love. With its fascinating exhibits and hands-on
experiences, the Museum of the Moving Image is a perfect destination
for any movie and TV aficionado.
Remember to plan your itinerary in a way that allows for travel
time between attractions, especially if you're relying on public
transportation or Uber. Be sure to check opening hours and any
reservation requirements for the attractions you wish to visit, as
they may vary. Enjoy your 24-hour adventure in the vibrant city of New
York!
```

Note that we've created the prompt in such a way that you can use it with different cities. Here's a run with London. Note that the response acknowledged my fear of heights, but simply told me to overcome it:

```
Attraction 1: Warner Bros. Studio Tour London - The Making of Harry
Potter
- What is the best thing to see there: Immerse yourself in the magical
world of Harry Potter by exploring the actual sets, costumes, and
props used in the film series.
Start your day by visiting the Warner Bros. Studio Tour London -
The Making of Harry Potter. As a fan of movies and TV shows, this
attraction will be a dream come true for you. Located just outside of
London, this studio tour offers a unique behind-the-scenes experience
of the Harry Potter film series.
Attraction 2: The British Museum
- What is the best thing to see there: Discover the vast collection of
art and artifacts from around the world, including the Rosetta Stone
and the Egyptian mummies.
Next, head to The British Museum, located in the heart of downtown
London. This world-renowned museum houses a vast collection of art
and historical artifacts, spanning over two million years of human
history. As a lover of movies and TV shows, you'll appreciate the
opportunity to see real-life artifacts and treasures from various
civilizations.
Attraction 3: The London Eye
- What is the best thing to see there: Overcome your fear of heights
and enjoy breathtaking panoramic views of London's iconic landmarks.
To end your day, challenge yourself and conquer your fear of heights
by visiting the London Eye. This giant Ferris wheel, located on
the South Bank of the River Thames, offers spectacular views of the
city's skyline. Take a ride in one of the glass capsules and enjoy a
30-minute rotation, offering uninterrupted 360-degree views of London.

With 24 hours in London, these three attractions will provide you with
a well-rounded experience that combines your love for movies and TV
shows with iconic landmarks and breathtaking views. Don't forget to
make use of public transportation or Uber to efficiently move between
these attractions and make the most of your limited time in the city.
Enjoy your adventure!
```

For many applications, you'll need to make your prompt more configurable, and that will require adding more variables. We'll do that next.

Prompts with multiple variables

You can parameterize a lot of information in a prompt. For example, in our prompt that finds the best attractions, you can add multiple variables, such as the number of days that the person is staying, the things they like and dislike, and how many attractions they want to see.

In such a case, our prompt will become more complex, so we will need to create a new `skprompt.txt` file. Since our prompt will be a new function and have multiple parameters, we will also need to create a new `config.json` file.

These two files can be found in the `plugins/prompt_engineering/attractions_multiple_variables` folder.

Skprompt.txt

To add more variables to a prompt, simply add them within double curly brackets with a dollar sign before their name. The following code shows how to add several variables (`city`, `n_days`, `likes`, `dislikes`, and `n_attractions`) to a single prompt:

```
I am traveling to {{$city}} and staying {{$n_days}} days.
I like {{$likes}}.
I dislike {{$dislikes}}.
I am traveling by myself.
Create an itinerary of up to {{$n_attractions}} must-see attractions
for me to visit.

The attractions should be listed in the following format:

- Attraction name: [name]
- What is the best thing to see there: [one sentence description]
Your whole answer should be less than 300 words.
```

Now, let's see the changes in the function configuration.

Config.json

Our `config.json` file for multiple variables is almost the same as the one we use with a single variable, but we need to add the details for all the variables:

```
    "input_variables": [
        {
            "name": "city",
            "description": "The city the person wants to travel to",
            "required": true
        },
        {
            "name": "n_days",
            "description": "The number of days the person will be in
the city",
            "required": true
        },
        {
            "name": "likes",
            "description": "The interests of the person traveling to
the city",
```

```
        "required": true
    },
    {

        "name": "dislikes",
        "description": "The dislikes of the person traveling to
the city",
        "required": true
    },
    {

        "name": "n_attractions",
        "description": "The number of attractions to recommend",
        "required": true

    }

]
```

Now that we've configured the new function, let's learn how to call it in Python and C#.

Requesting a complex itinerary with Python

Compared to calling a prompt with a single parameter, the only change we need to make to call a prompt with multiple parameters is on the KernelArguments object. When passing KernelArguments as a parameter to kernel.invoke, we must add all the parameters we need to the object, as shown here. One thing to note is that the parameters are all strings since LLMs work best with text:

```
    response = await kernel.invoke(pe_plugin["attractions_
multiple_variables"], KernelArguments(
    city = "New York City",
    n_days = "3",
    likes = "restaurants, Ghostbusters, Friends tv show",
    dislikes = "museums, parks",
    n_attractions = "5"
))
```

Let's see the C# code.

Requesting a complex itinerary with C#

This time, we will create a KernelArguments object called function_arguments outside of the function invocation and pre-fill the five variables with the content we want. Then, we will pass this object to the invocation call:

```
var function_arguments = new KernelArguments()
    {["city"] = "New York City",
    ["n_days"] = "3",
```

```
    ["likes"] = "restaurants, Ghostbusters, Friends tv show",
    ["dislikes"] = "museums, parks",
    ["n_attractions"] = "5"};

var result = await kernel.InvokeAsync(promptPlugin["attractions_
multiple_variables"], function_arguments );
```

Now, let's see the results.

The result of the complex itinerary

The result considers all the input variables. It suggests a park – the High Line – even though I told the LLM that I dislike parks. It does explain that it is a very unusual park, and if you know New York, it doesn't feel like a park at all. I think that a person who doesn't enjoy the traditional park experience would enjoy the High Line, so the LLM did a very good job once it got more context:

```
Itinerary for Three Days in New York City:
Day 1:
1. Attraction name: Ghostbusters Firehouse
    What is the best thing to see there: Visit the iconic firehouse
featured in the Ghostbusters movies and take pictures in front of the
famous logo.

2. Attraction name: Friends Apartment Building
    What is the best thing to see there: Pay a visit to the apartment
building that served as the exterior shot for Monica and Rachel's
apartment in the beloved TV show Friends.
Day 2:
3. Attraction name: Restaurant Row
    What is the best thing to see there: Explore Restaurant Row on West
46th Street, known for its diverse culinary scene, offering a plethora
of international cuisines to satisfy your food cravings.
4. Attraction name: High Line Park
    What is the best thing to see there: Although you mentioned
disliking parks, the High Line is a unique urban park built on a
historic freight rail line, offering beautiful views of the city and a
different experience compared to traditional parks.

Day 3:
5. Attraction name: Times Square
    What is the best thing to see there: Immerse yourself in the
vibrant atmosphere of Times Square, known for its dazzling billboards,
bustling streets, and renowned theaters, making it a must-see
destination in NYC.
This itinerary focuses on your interests while also incorporating some
iconic NYC experiences. You can explore famous film locations like the
Ghostbusters Firehouse and Friends Apartment Building. Since you enjoy
```

```
restaurants, Restaurant Row will offer a variety of dining options
to suit your taste. Although you specified disliking parks, the High
Line Park provides a unique urban green space experience. Finally, no
trip to NYC would be complete without a visit to the energetic Times
Square. Enjoy your trip!
```

With that, we've learned how to improve our prompts to get better results. However, there are some cases in which LLMs fail to provide good answers, even after we use all the techniques mentioned previously. A very common case is solving math problems. We will explore this next.

Issues when answering math problems

Although the results from LLMs are impressive, sometimes, models can get confused by seemingly simple questions. This tends to be more frequent when the question involves math. For example, I ran the following prompt in GPT-3.5 five times:

```
When I was 6 my sister was half my age.
Now I'm 70. How old is my sister?
```

The plugin I used to run this prompt is under `plugins/prompt_engineering/solve_ math_problem`.

The correct answer is 67 because when I was 6, my sister was 3. Now, 64 years later, I'm 70, so she would be 67.

Here are the results of the five runs on GPT-3.5. The first result was incorrect, saying that "*my sister is 64 years younger than her current age:*"

```
If your sister was half your age when you were 6, that means she was 3
years younger than you at that time (6/2 = 3).
Since then, you have aged 64 years (70 - 6 = 64). Therefore, your
sister would also be 64 years younger than her current age, which
would make her 70 - 64 = 6 years old.
So, your sister is currently 6 years old.
```

The second and third attempts gave the correct answer. Here's the second attempt:

```
If your sister was half your age when you were 6, it means she was 6/2
= <<6/2=3>>3 years old at that time.
Since then, there is a constant age difference of 3 years between the
two of you.
Now that you are 70, your sister would be 70 - 3 = <<70-3=67>>67 years
old.
```

Here's the third attempt:

```
If your sister was half your age when you were 6, that means she was 3
years old at that time.
Since then, the age difference between you and your sister remains
constant.
Therefore, if you are currently 70 years old, your sister would be 67
years old.
```

But the fourth and fifth attempts gave wrong answers again. Here's the fourth attempt:

```
If your sister was half your age when you were 6, that means she was 3
years old at that time. Since then, the age difference between you and
your sister remains constant. Therefore, if you are currently 70 years
old, your sister is 73 years old.
```

Here's the fifth attempt:

```
If your sister was half your age when you were 6, that means she was 3
years younger than you at that time.
Now, you are 70, which means there is a 67-year age difference between
when you were 6 and now.
So, if your sister was 3 years younger than you when you were 6, she
would be 67 years younger than you now.
Therefore, your sister would be 70 - 67 = 3 years old.
```

Therefore, in five attempts, only two (40%) returned correct answers, which is not an experience that you want your users to have.

We will learn how to address this issue in the next section.

Multistage prompts

One way to improve the accuracy of LLMs when doing math is to use multistage prompts. In this technique, the answer from the first prompt is passed to the second as a parameter. We're going to illustrate this with the **Chain-of-Thought (CoT)** technique.

CoT – "Let's think step by step"

In the paper *Large Language Models are Zero-Shot Reasoners* [1], the authors found that simply adding "Let's think step by step" right after the question can help improve the accuracy of LLMs a lot. Their proposed process works as follows:

1. Ask the intended question, but instead of asking the LLM to answer, simply append "*Let's think step by step*" at the end.

2. The LLM will answer with a process to answer the question.

3.   Combine the question from *step 1* with the process from *step 2* in a new prompt, and finish with "*Therefore, the answer is…:*"

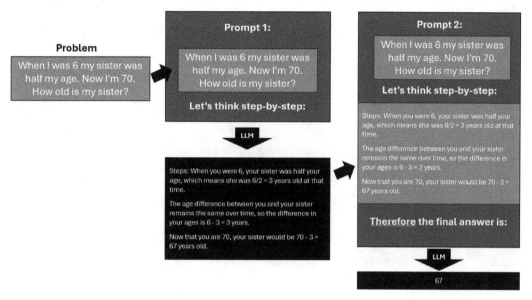

Figure 2.1 – The Zero-shot-CoT method

They called their process *Zero-shot-Chain-of-Thought* or *Zero-shot-CoT*. The *Zero-shot* part means that you don't need to give any example answers to the LLM; you can give the question directly. This is to differentiate the process from **few-shot**, which is when you provide the LLM several examples of expected answers in your prompt, making the prompt substantially larger. The *CoT* part describes the process of asking the LLM to provide a reasoning framework and adding the LLM's reasoning to the question.

The authors tested several different phrases to obtain the CoT from the LLM, such as "*Let's think about it logically*" and "*Let's think like a detective step by step,*" and found that simply "*Let's think step by step*" yielded the best results.

Implementing Zero-shot-CoT

We need two prompts – one for both steps. For the first step, we will call `solve_math_problem_v2`. The prompt simply restates the problem and adds "*Let's think step by step*" at the end:

```
{{$problem}}
Let's think step by step.
```

The prompt for the second step, which we will call `chain_of_thought`, repeats the first prompt, includes the answer for the first prompt, and then asks for the solution:

```
{{$problem}}
Let's think step by step.
{{$input}}
Therefore, the answer is…
```

Given this prompt, the `config.json` file needs two `input_variables`:

```
"input_variables": [
    {
        "name": "problem",
        "description": "The problem that needs to be solved",
        "required": true
    },
    {
        "name": "input",
        "description": "The steps to solve the problem",
        "required": true
    }
]
```

Let's learn how to call the prompts.

Implementing CoT with Python

To make the steps explicit, I broke the program into two parts. The first one asks for the CoT and shows it:

```
problem = """When I was 6 my sister was half my age. Now I'm 70.
How old is my sister?"""

pe_plugin = kernel.add_plugin(None, parent_directory="../../
plugins", plugin_name="prompt_engineering")
solve_steps = await kernel.invoke(pe_plugin["solve_math_problem_
v2"], KernelArguments(problem = problem))
print(f"\n\nSteps: {str(solve_steps)}\n\n")
```

The second part shows the answer. If all you care about is the answer, you don't need to print the steps, but you still need to use the LLM to calculate the steps because they are a required parameter to the CoT:

```
response = await kernel.invoke(pe_plugin["chain_of_thought"],
KernelArguments(problem = problem, input = solve_steps))
print(f"\n\nFinal answer: {str(response)}\n\n")
```

Implementing CoT with C#

As we did in Python, we can break the C# program into two parts. Part one will show the reasoning steps elicited by the CoT prompt:

```
var problem = "When I was 6 my sister was half my age. Now I'm 70. How
old is my sister?";

var chatFunctionVariables1 = new KernelArguments()
{
    ["problem"] = problem,
};

var steps = await kernel.InvokeAsync(promptPlugin["solve_math_problem_
v2"], chatFunctionVariables1);
```

The second part parses and reports the answer:

```
var chatFunctionVariables2 = new KernelArguments()
{
    ["problem"] = problem,
    ["input"] = steps.ToString()
};

var result = await kernel.InvokeAsync(promptPlugin["chain_of_thought_
v2"], chatFunctionVariables2);Console.WriteLine(steps);
```

Let's see the results.

Results for CoT

I ran the program five times, and it got the right answer every time. The answers aren't deterministic, so it may be that you run it on your machine and get one or two wrong answers. In the paper, the authors claimed that this method achieves 78.7% success in problems of this type, where the usual accuracy of LLMs is around 17.7%.

Let's look at two sample responses. Here's the first:

```
Steps: When you were 6, your sister was half your age, which means she
was 6/2 = 3 years old at that time.
The age difference between you and your sister remains the same over
time, so the difference in your ages is 6 - 3 = 3 years.
Now that you are 70, your sister would be 70 - 3 = 67 years old.
Final answer:
Your sister is 67 years old.
```

Here's the second:

```
Steps: When you were 6, your sister was half your age, so she was 6/2
= <<6/2=3>>3 years old.
Since then, the age difference between you and your sister remains
constant, so your sister is 3 years younger than you.
Therefore, if you are now 70, your sister would be 70 - 3 = <<70-
3=67>>67 years old.
```

Instead of running the program manually, we can automate it. Let's see how.

An ensemble of answers

While the CoT technique helps a lot, we still have a 78.7% average accuracy, and that may not be enough. To address this problem, one technique that is frequently used is to ask the model the same question several times and compile only the most frequently given answer.

To achieve this, we will make a minor modification to our CoT prompt and call it `chain_of_thought_v2`. We will simply ask the LLM to answer in Arabic numerals, to make it easier to compare the answers in a later step:

```
{{$problem}}
Let's think step by step.
{{$input}}
Using arabic numerals only, the answer is…
```

We also need to change the program and ask it to run several times. For the next example, I chose $N = 7$. We will collect the answers and choose the answer that appears more frequently. Note that each call using this method is N times more expensive and takes N times longer than a single call. Accuracy is not free.

Automatically running an ensemble with Python

Let's run CoT seven times. For each time we run it, we'll add the result to a list. Then, we'll take advantage of the `set` data structure to quickly get the most common element:

```
responses = []

for i in range(7):
    solve_steps = await kernel.invoke(pe_plugin["solve_math_
problem_v2"], KernelArguments(problem = problem))
    response = await kernel.invoke(pe_plugin["chain_of_thought_
v2"], KernelArguments(problem = problem, input = solve_steps))
    responses.append(int(str(response)))

print("Responses:")
```

```
        print(responses)

        final_answer = max(set(responses), key = responses.count)
        print(f"Final answer: {final_answer}")
```

Let's see how to implement this in C#.

Automatically running an ensemble with C#

The C# code uses the same idea as the Python code: we run the model seven times and store the results. Then, we search the results for the most frequent answer:

```
var results = new List<int>();

for (int i = 0; i < 7; i++)
{
    var chatFunctionVariables1 = new KernelArguments()
    {
        ["problem"] = problem,
    };

    var steps = await kernel.InvokeAsync(promptPlugin["solve_math_
problem_v2"], chatFunctionVariables1);

    var chatFunctionVariables2 = new KernelArguments()
    {
        ["problem"] = problem,
        ["input"] = steps.ToString()
    };

    var result = await kernel.InvokeAsync(promptPlugin["chain_of_
thought_v2"], chatFunctionVariables2);

    var resultInt = int.Parse(result.ToString());

    results.Add(resultInt);
}

var mostCommonResult = results.GroupBy(x => x)
    .OrderByDescending(x => x.Count())
    .First()
    .Key;

Console.WriteLine($"Your sister's age is {mostCommonResult}");
```

Combining CoT and the ensemble method substantially increases the likelihood of getting a correct response. In the paper, the authors obtained 99.8% correct results, at the expense of making 10 LLM calls per question.

Summary

In this chapter, you learned several techniques you can utilize to improve your prompts to obtain better results. You learned to use longer prompts that ensure that the LLM has the necessary context to provide the desired response. You also learned how to add multiple parameters to a prompt. Then, you learned how to chain prompts and how to implement the CoT method to help the LLM provide more accurate results. Finally, you learned how to ensemble several responses to increase accuracy. This accuracy, however, comes at a cost.

Now that we have mastered prompts, in the next chapter, we will explore how to customize plugins and their native and semantic functions.

References

[1] T. Kojima, S. S. Gu, M. Reid, Y. Matsuo, and Y. Iwasawa, "Large Language Models are Zero-Shot Reasoners." arXiv, Jan. 29, 2023. Accessed: Jun. 06, 2023. [Online]. Available: `http://arxiv.org/abs/2205.11916`

Part 2:
Creating AI Applications
with Semantic Kernel

In this part, we will go deep inside Semantic Kernel and learn how to use it to solve problems. We start by adding functions to a kernel, and then we use functions to solve a problem. The real power comes next when we ask the kernel to solve a problem on its own. Finally, we learn how to keep history for our kernel using memory.

This part includes the following chapters:

- *Chapter 3, Extending Semantic Kernel*
- *Chapter 4, Performing Complex Actions by Chaining Functions*
- *Chapter 5, Programming with Planners*
- *Chapter 6, Adding Memories to Your AI Application*

3

Extending Semantic Kernel

Plugins encapsulate AI capabilities into a single unit of functionality. Plugins are the building blocks of AI applications. We will learn about the different types of functions (native functions and semantic functions) and how to expose these functions as a plugin, so that they can be used by Semantic Kernel.

In previous chapters, we created and used very simple plugins. In this chapter, we will start by describing the pre-existing tools that you already have available just by installing Semantic Kernel. Then, we will explore the best practices for developing your own **native plugins**, which are collections of native functions. These functions perform tasks that AI cannot do well, such as querying a database or doing complex math. Lastly, you will learn how to create **semantic plugins**, which are collections of semantic functions. These functions are executed by an AI service such as OpenAI's GPT. We will explore the different parameters and how they work together. Throughout this chapter, we will use an application that evaluates grant requests for a nonprofit as our use case.

In this chapter, we'll be covering the following topics:

- The out-of-the-box plugins
- Creating native functions
- Creating semantic functions

By the end of the chapter, you will have the tools required to build a kernel that can perform many coordinated functions, and that will be easier to use as a copilot or to be integrated into other tools such as the planner. In the next chapter, we will use this kernel to process a pipeline of documents.

Technical requirements

To complete this chapter, you will need to have a recent, supported version of your preferred Python or C# development environment:

- For Python, the minimum supported version is Python 3.10, and the recommended version is Python 3.11
- For C#, the minimum supported version is .NET 8

In this chapter, we will call OpenAI services. Given the amount that companies spend on training these LLMs, it's no surprise that using these services is not free. You will need an **OpenAI API** key, obtained either directly through **OpenAI** or **Microsoft**, via the **Azure OpenAI** service.

If you are using .NET, the code for this chapter is at `https://github.com/PacktPublishing/Building-AI-Applications-with-Microsoft-Semantic-Kernel/tree/main/dotnet/ch3`.

If you are using Python, the code for this chapter is at `https://github.com/PacktPublishing/Building-AI-Applications-with-Microsoft-Semantic-Kernel/tree/main/python/ch3`.

You can install the required packages by going to the GitHub repository and using the following: `pip install -r requirements.txt`.

Getting to know the core plugins

Large language models are great at generating text and images, but they are currently unreliable for several other tasks, for example:

- Retrieving data from external data sources, such as the filesystem and the web
- Interacting with the system's clock
- Performing math, even simple arithmetic functions

The Semantic Kernel installation includes several plugins for tasks that are very frequently used. The core plugins are native plugins that provide functionality that LLMs struggle with.

To use the plugins in C#, you need to import the `Microsoft.SemanticKernelCoreSkills` library, and to use the plugins in Python, you need to import the plugins from the `semantic_kernel.core_skills` library.

> **Skills or plugins?**
> When the Microsoft Semantic Kernel was first released, plugins were called *skills*. This changed when OpenAI allowed developers to build extensions to ChatGPT and called these extensions *plugins*. Since the purpose is the same, the Semantic Kernel team decided to change the name from *skills* to *plugins*, but the old name lives on in the existing code in some places. This is likely to change in the future.

The following list shows the plugins that are currently available in Semantic Kernel:

Plugin Name	What it does	Available in Python?	Available in C#?
TimePlugin	Acquires the time of day and any other temporal information	✓	✓
ConversationSummaryPlugin	Summarizes a conversation	✓	✓
FileIOPlugin	Reads and writes to the filesystem	✓	✓
HttpPlugin	Calls APIs	✓	✓
MathPlugin	Performs mathematical operations	✓	✓
TextMemoryPlugin	Stores and retrieves text in memory	✓	✓
TextPlugin	Manipulates text strings deterministically	✓	✓
WaitPlugin	Pauses execution for a specified amount of time	✓	✓
WebSearchPlugin	Performs Bing web searches and returns results	✓	✗

Table 3.1 – Core plugins and their functions

At the time of writing , the core plugins don't have documentation. To see their parameters, you must read their code on the GitHub repository. For example, to find the parameters for the TimePlugin, you have to go directly to the repository:

- C#: https://github.com/microsoft/semantic-kernel/blob/main/dotnet/src/Plugins/Plugins.Core/TimePlugin.cs

- Python: https://github.com/microsoft/semantic-kernel/blob/main/python/semantic_kernel/core_plugins/time_plugin.py

Here are the parameters available for TimePlugin:

```
{{time.date}}          => Sunday, 12 January, 2031
{{time.today}}         => Sunday, 12 January, 2031
{{time.iso_date}}      => 2031-01-12
{{time.now}}           => Sunday, January 12, 2031 9:15 PM
{{time.utcNow}}        => Sunday, January 13, 2031 5:15 AM
{{time.time}}          => 09:15:07 PM
{{time.year}}          => 2031
```

```
{{time.month}}              => January
{{time.monthNumber}}        => 01
{{time.day}}                => 12
{{time.dayOfWeek}}          => Sunday
{{time.hour}}               => 9 PM
{{time.hourNumber}}         => 21
{{time.days_ago $days}}     => Sunday, 7 May, 2023
{{time.last_matching_day $dayName}} => Sunday, 7 May, 2023
{{time.minute}}             => 15
{{time.minutes}}            => 15
{{time.second}}             => 7
{{time.seconds}}            => 7
{{time.timeZoneOffset}}     => -0800
{{time.timeZoneName}}       => PST
```

Now that we know the available parameters, let's see a quick sample of what the plugin can do.

To use the out-of-the-box plugins in C#, you first need to install the `Plugins.Core` package. To install it, use the following command (change the version to match your Semantic Kernel version):

```
dotnet add package Microsoft.SemanticKernel.Plugins.Core --version
1.7.1-alpha
```

An example – Using the TimePlugin

We're going to use the `TimePlugin` as part of the prompt of a semantic function. First, we create an instance of the `TimePlugin` and name it `time`.

Creating the TimePlugin in C#

The core plugins for C# are still in pre-release. To use them, you need to add a #`pragma` directive (`SKEXP0050`) to disable the pre-release warning. You also need to import the `Microsoft.SemanticKernel.Plugins.Core` package.

We are going to add the plugin using the `AddFromType` method of the Semantic `KernelBuilder` object:

```
# pragma warning disable SKEXP0050

using Microsoft.SemanticKernel;
using Microsoft.SemanticKernel.Plugins.Core;

var (apiKey, orgId) = Settings.LoadFromFile();

var builder = Kernel.CreateBuilder()
                    .AddOpenAIChatCompletion("gpt-3.5-turbo",
```

```
    apiKey, orgId, serviceId: "gpt35");
builder.Plugins.AddFromType<TimePlugin>(pluginName: "time");
Kernel kernel = builder.Build();
```

Let's see how to do it in Python next.

Creating the TimePlugin in Python

In Python, the core plugins are in the `semantic_kernel.core_plugins` package. Once you import the plugin you want to use from that package, you can load it into the kernel with the `add_plugin` method:

```
import asyncio
from semantic_kernel.connectors.ai.open_ai import OpenAIChatCompletion
from semantic_kernel.utils.settings import openai_settings_from_dot_
env
import semantic_kernel as sk
from semantic_kernel.core_plugins.time_plugin import TimePlugin

async def main():

    kernel = sk.Kernel()
    api_key, org_id = openai_settings_from_dot_env()
    gpt35 = OpenAIChatCompletion("gpt-3.5-turbo", api_key, org_id,
"gpt35")

    kernel.add_service(gpt35)
    kernel.add_plugin(TimePlugin(), "time")
```

Then, we create a semantic function and invoke it, as we did in both previous chapters:

Invoking the TimePlugin in C#

```
const string promptTemplate = @"
Today is: {{time.date}}
Current time is: {{time.time}}

Answer to the following questions using JSON syntax, including the
data used.
Is it morning, afternoon, evening, or night (morning/afternoon/
evening/night)?
Is it weekend time (weekend/not weekend)?";

var results = await kernel.InvokePromptAsync(promptTemplate);
Console.WriteLine(results);
```

Invoking the TimePlugin in Python

```
prompt = """
Today is: {{time.date}}
Current time is: {{time.time}}

Answer to the following questions using JSON syntax, including the
data used.
Is it morning, afternoon, evening, or night (morning/afternoon/
evening/night)?
Is it weekend time (weekend/not weekend)?
"""
prompt_function = kernel.add_function(function_name="ex03",
plugin_name="sample", prompt=prompt)
response = await kernel.invoke(prompt_function, request=prompt)
```

Results

For both languages, you should expect to get a JSON file that gives you information about the current time. Since we did not specify the exact JSON format in our prompt, the fields chosen by your invocation may be different:

```
{
    "date": "Wednesday, 21 June, 2023",
    "time": "12:17:02 AM",
    "period": "night",
    "weekend": "not weekend"
}
```

While there's a lot that you can do with the core plugins, in most cases you'll have to develop your own plugins. In the remainder of this chapter, we will explore the details of developing native and semantic functions.

But first, let's look at our sample application that we'll be employing to study the development of native and semantic plugins.

Introducing the application – Validating grants

For this chapter and the next, imagine you are working for an organization that gives monetary grants to projects. For example, this could be the Department of Education giving away $100,000 to projects that will teach AI in high schools, or a non-profit organization such as the Gates Foundation giving $1,000,000 away to the three best proposals to eradicate malaria. The same concept also applies to several processes in corporations: for example, before sending a quote to a customer, a corporation may require that it fulfills some requirements, such as having a price, an expiration date, and some legal language. Another example is submitting visa applications in the United States. Lawyers must send several documents documenting the candidate's academic and professional skills, in addition to several forms. If they miss one form or document, the whole application is rejected and they have to submit it again, including having to pay all the fees again.

In these next two chapters, we are going to write a simple application that evaluates a Word document that contains a grant request for a vaccination project and an Excel file with a budget. Our program will evaluate whether it fulfills the requirements of our grant program. If it doesn't, it will tell the submitter what's missing. To achieve that, we will write a native plugin to evaluate the Excel file, another native plugin to break the Word document into parts, and a semantic plugin to evaluate the contents of the Word document that describes project proposal.

> **The second random selection for H-1B visas in 2021**
>
> In July of 2021, the **U.S. Citizenship and Immigration Services** (**USCIS**) determined that they needed to conduct a second round of H-1B applications for their employment-based visa. The H-1B visa is used mostly by tech companies, and some large tech companies such as Amazon, Alphabet, Meta, and Microsoft had to file tens of thousands of applications in a short period of time. The applications needed to be manually reviewed by lawyers, an expensive and time-consuming project. Any error in the application would cause it to be automatically rejected, which could result in people having to leave the United States. The techniques we show in these two chapters can be used to build an application that reviews thousands of documents in minutes. Manual reviews are still required, but at least the most egregious errors will be automatically detected, saving the time of lawyers.

Directory structure of our application

In our example in this and the following chapter, we will create an application that evaluates grant requests for a vaccination program by checking the content of two documents: a Word document and an Excel spreadsheet.

Each grant request will be in a directory containing the Excel spreadsheet that has the budget requested for the vaccination campaign and the Word document that describes the team that will be executing the vaccination program and their experience.

The data is available in the data directory of the book's GitHub repository, and its directory structure is as follows:

```
data
├───correct
│       correct.docx
│       correct.xlsx
├───incorrect01
│       incorrect_template.xlsx
│       missing_experience.docx
```

```
├──incorrect02
|       missing_qualifications.docx
|       over_budget.xlsx
├──incorrect03
|       correct.docx
|       fast_increase.xlsx
├──incorrect04
|       correct.docx
|       incorrect_cells.xlsx
├──incorrect05
|       incorrect_cells.xlsx
├──incorrect06
|       correct.docx
├──incorrect07
|       correct.xlsx
|       missing_experience.docx
├──incorrect08
|       correct.xlsx
|       missing_qualifications.docx
├──incorrect09
|       correct.xlsx
|       wrong_dates.docx
└──incorrect10
        correct.docx
        correct.xlsx
        missing_experience.docx
```

For this chapter, we will use the first four grant requests. One set of documents is valid, which is stored in the directory named `correct`, while the other three, which are stored in `incorrect1`, `incorrect2` and `incorrect3`, respectively, have problems, which will be described later in this chapter.

Evaluating the *structure* of documents is something that is easier and more reliable with regular programming than with LLMs. On the other hand, evaluating the *content* of documents is something that is easier with LLMs than with regular programming.

We are going to do the first part using native functions.

Developing native plugins

Let's start with native functions. Native functions are regular code in your language of choice (Python or C#) and do not necessarily require a specific directory structure, unlike semantic functions. However, it is easier to put the native functions together with the main code of your program. This will make it easier to import the classes.

> **Avoid duplicate names**
>
> Since semantic functions are directories and native functions are source code files, it's possible to have a semantic function and a native function with the same name inside a plugin, for example a directory named `my_function` and a source code file named `my_function.py`. If you do that, the last function loaded (the source code file) will overwrite the first, leading to unexpected problems. Therefore, avoid duplicate names. At the time of writing, Semantic Kernel does not provide a warning when this happens.

Let's look at the directory structure of our plugins.

The directory structure of our plugins

Semantic Kernel offers a function to add plugins from a directory. While you can always load plugins one at a time, setting up the plugins in the following directory structure makes it easier to load them into the kernel with fewer function calls:

```
└──ch3
    └──proposals
    └──plugins
    │    └──ProposalChecker
    │          ├──CheckQualifications
    │          │      ├──skprompt.txt
    │          │      └──config.json
    │          ├──CheckExperience
    │          │      ├──skprompt.txt
    │          │      └──config.json
    │          └──CheckImplementationDescription
    │                 ├──skprompt.txt
    │                 └──config.json
    └──code
         ├──python
         │    ├──program.py
         │    ├──ParseWordDocument.py
         │    └──CheckSpreadsheet.py
         └──dotnet
              ├──Program.cs
              ├──ParseWordDocument.cs
              └──CheckSpreadsheet.cs
```

The first three directories, `CheckQualifications`, `CheckExperience`, and `CheckImplementationDescription`, are semantic functions and will be addressed in the *Developing semantic plugins* section. The `CheckSpreadsheet` and `ParseWordDocument` files contain native functions that perform the following:

- `CheckSpreadsheet`: Validates the spreadsheet to make sure it contains all required fields
- `ParseWordDocument`: Validates the Word document and makes sure it contains all the required headings

Checking the structure of our Excel spreadsheet

The desired structure of the Excel spreadsheet for our application is as follows:

- The Excel file needs to have exactly two sheets, one named 2024 and another named 2025
- Each sheet needs to follow the following template:

Quarter	Budget
Q1	$99,999
Q2	$99,999
Q3	$99,999
Q4	$99,999

It needs to follow these rules:

- The total budget to implement the vaccination campaign for each year needs to be below $1,000,000
- No more than a 10% increase quarter over quarter

To test the function, we created the following Excel workbooks:

- `correct.xslx`: A workbook that follows all the rules and should pass the test
- `incorrect_template.xlsx`: A workbook that doesn't have the sheets named 2024 and 2025 and should fail the test
- `over_budget.xlsx`: A spreadsheet that follows the format but requests more than a million dollars per year and should fail the test
- `fast_increase.xlsx`: A spreadsheet that follows the format and is under budget, but has expenses increasing by more than 10% quarter over quarter, and should therefore fail the test

We will implement the code to check these spreadsheets in the following section.

Using native functions to check the Excel spreadsheet

For both languages, native functions must be defined as public methods in a class that represents your plugin. The functions will return text because that will be easier to integrate with semantic functions later.

Python

In Python, we're going to use the **OpenPyXL** package (https://openpyxl.readthedocs.io/en/stable/), which makes it easy to open an Excel file, ensure it has the required number of tabs, ensure that it has the required tables, and load the tables into variables:

```
import openpyxl
from typing_extensions import Annotated
from semantic_kernel.functions.kernel_function_decorator import
kernel_function

class CheckSpreadsheet:
```

C#

```
using System.ComponentModel;
using Microsoft.SemanticKernel;
using OfficeOpenXml;

namespace Plugins.ProposalChecker;
public class CheckSpreadsheet
{
}
```

Now that we have the class templates created, we will need to add the functions that implement the functionality that checks the spreadsheet.

To add functions to your class, you need to use the KernelFunction (C#) or kernel_function (Python) decorator. This will enable Semantic Kernel to recognize the function. When using the decorator, you need to add the Description attribute for the function and its parameters. This description can be used by planners to decide which functions to use when the user sends a request, as we saw in *Chapter 1*.

Our first function, `CheckTabs`, checks that the spreadsheet has the two required tabs named 2024 and 2025:

Python

```
    @kernel_function(
        description="Checks that the spreadsheet contains the correct
tabs, 2024 and 2025",
        name="CheckTabs",
    )
    def CheckTabs(self,
                   path: Annotated[str, "The path to the spreadsheet"])
-> Annotated[str, "The result of the check"]:
        try:
            workbook = openpyxl.load_workbook(path)
            sheet_names = workbook.sheetnames
            if sheet_names == ['2024', '2025']:
                return "Pass"
            else:
                return "Fail: the spreadsheet does not contain the
correct tabs"
        except Exception as e:
            return f"Fail: an exception {e} occurred when trying to
open the spreadsheet"
```

The function opens the workbook with the `load_workbook` method of `openpyxl` and checks that it has exactly two tabs named 2024 and 2025. If it succeeds, it returns `"Pass"`. Otherwise, it returns `"Fail"` and provides an error description.

In Python, you can tell the name you want Semantic Kernel to use with the function by using the name attribute of the `kernel_function` decorator. You can also describe your input parameters using the `Annotated` decorators. The first parameter of the `Annotated` decorator is the type of the input parameter and the second is its description.

C#

```
    [KernelFunction, Description("Checks that the spreadsheet contains
the correct tabs, 2024 and 2025")]
    public string CheckTabs([Description("The file path to the
spreadsheet")] string filePath)
    {
        try
        {
            FileInfo fileInfo = new FileInfo(filePath);

            if (!fileInfo.Exists)
```

```
        {
            return "Fail: File does not exist.";
        }
        using (var package = new ExcelPackage(fileInfo))
        {
            ExcelPackage.LicenseContext = OfficeOpenXml.
LicenseContext.NonCommercial;
            var workbook = package.Workbook;
            if (workbook.Worksheets.Count != 2)
            {
                return "Fail: Spreadsheet does not contain 2
tabs.";
            }
            if (workbook.Worksheets.Any(sheet => sheet.Name ==
"2024") && workbook.Worksheets.Any(sheet => sheet.Name == "2025"))
            {
                return "Pass";
            }
            else
            {
                return "Fail: Spreadsheet does not contain 2024
and 2025 tabs.";
            }
        }
    }
    catch (Exception ex)
    {
        return $"Fail: An error occurred: {ex.Message}";
    }
}
```

The function opens the workbook with the `ExcelPackage` class of `OfficeOpenXml` and checks that it has exactly two tabs named `2024` and `2025`. If it succeeds, it returns `"Pass"`. Otherwise, it returns `"Fail"` and provides an error description.

In C#, we use parameter decorators to add a `Description` attribute to parameters when declaring a function, and a `KernelFunction` function decorator that makes the function accessible to `SemanticKernel` and describes its purpose with the `Description` attribute.

Now that we have the plugin and the test data set up, let's write a simple program that will perform just the first two tests: checking that the spreadsheet exists and checking whether it contains the tabs we expect. We will add more tests later.

Checking the contents of the worksheets

In the following code, we are going to check the contents of each sheet:

C#

```
using Microsoft.SemanticKernel;
using System.IO;
var (apiKey, orgId) = Settings.LoadFromFile();

Kernel kernel = Kernel.CreateBuilder()
                      .AddOpenAIChatCompletion("gpt-3.5-turbo",
apiKey, orgId, serviceId: "gpt35")
                      .Build();

var checkerPlugin = kernel.ImportPluginFromObject(new Plugins.
ProposalChecker.CheckSpreadsheet());
```

Using the familiar template, we use the `Settings.LoadFromFile` method to load the API key, use `KernelBuilder` and `AddOpenAIChatCompletion` to connect to the OpenAI service for later, and then use the `ImportPluginFromObject` method of the kernel to import the `CheckSpreadsheet` class.

Now we can check the spreadsheets by using `InvokeAsync` and passing the `CheckTabs` function as its function parameter, and each file as its `filePath` parameter:

```
// Check for tabs
var result1 = await kernel.InvokeAsync(checkerPlugin["CheckTabs"],
new() {["filePath"] = $"{data_directory}/correct/correct.xlsx"});
var result2 = await kernel.InvokeAsync(checkerPlugin["CheckTabs"],
new() {["filePath"] = $"{data_directory}/incorrect01/incorrect_
template.xlsx"});
var result3 = await kernel.InvokeAsync(checkerPlugin["CheckTabs"],
new() {["filePath"] = $"{data_directory}/incorrect02/over_budget.
xlsx"});
var result4 = await kernel.InvokeAsync(checkerPlugin["CheckTabs"],
new() {["filePath"] = $"{data_directory}/incorrect03/fast_increase.
xlsx"});

Console.WriteLine("Checking whether the correct tabs are present in
the spreadsheet:");
Console.WriteLine(result1);
Console.WriteLine(result2);
Console.WriteLine(result3);
Console.WriteLine(result4);
```

Let's see how to perform the checks with Python:

Python

```python
import asyncio
from semantic_kernel.connectors.ai.open_ai import OpenAIChatCompletion
import semantic_kernel as sk
from semantic_kernel.utils.settings import openai_settings_from_dot_
env
from CheckSpreadsheet import CheckSpreadsheet
from ParseWordDocument import ParseWordDocument
from semantic_kernel.functions.kernel_arguments import KernelArguments

async def run_spreadsheet_check(path, function):
    kernel = sk.Kernel()

    check_spreadsheet = kernel.add_plugin(CheckSpreadsheet(),
"CheckSpreadsheet")

    result = await kernel.invoke(
        check_spreadsheet[function], KernelArguments(path = path)
    )
    print(result)
async def main():
    data_path = "../../data/proposals"
    await run_spreadsheet_check(f"{data_path}/correct/correct.xlsx",
"CheckTabs")
    await run_spreadsheet_check(f"{data_path}/incorrect01/incorrect_
template.xlsx", "CheckTabs")
    await run_spreadsheet_check(f"{data_path}/incorrect02/over_budget.
xlsx", "CheckTabs")
    await run_spreadsheet_check(f"{data_path}/incorrect03/fast_
increase.xlsx", "CheckTabs")
# Run the main function
if __name__ == "__main__":
    asyncio.run(main())
```

In Python, we use the `add_plugin` method to import the native function class and the `kernel.invoke` method to invoke it. In the preceding code, I put the functionality in a function called `run_spreadsheet_check`.

Results

We are running the test on four files, one that we expect to pass and three that we expect to fail. The results come out as expected:

```
Pass
Fail: the spreadsheet does not contain the correct tabs
Fail: Sum of values in year 2025 exceeds 1,000,000.
Fail: More than 10% growth found from B2 to B3 in sheet 2024.
```

Now that we have our initial template done, we need to add the other functions to the plugin to perform all the checks we need.

Additional checks

We still need to check the following:

- Whether there are exactly 10 cells filled in each tab
- Whether the budget adds up to less than $1,000,000
- Whether the budget grows by 10% or less each quarter

Let's start by checking whether the spreadsheet has all the cells we expect:

Python

In Python, we create a dictionary with the list of expected cells and make sure that they are filled. Don't forget to use the @kernel_function decorator, otherwise Semantic Kernel won't be able to find the function:

```python
    @kernel_function(
        description="Checks that the spreadsheet contains the correct
cells A1-B5",
        name="CheckCells",
    )
    def CheckCells(self,
                path: Annotated[str, "The path to the spreadsheet"])
-> Annotated[str, "The result of the check"]:
        workbook = openpyxl.load_workbook(path)
        required_cells = {
            'A1': 'Quarter', 'B1': 'Budget',
            'A2': 'Q1', 'A3': 'Q2', 'A4': 'Q3', 'A5': 'Q4'
        }
        for year in ['2024', '2025']:
            sheet = workbook[year]
```

```
                    for cell, value in required_cells.items():
                        if sheet[cell].value != value:
                            return "Fail: missing quarters"
                    for row in range(2, 6):
                        if not isinstance(sheet[f'B{row}'].value, (int,
float)):
                            return "Fail: non-numeric inputs"
            return "Pass"
```

C#

In C#, we will check the cells one by one. It is still a very straightforward function, where we simply look at each cell and check to see whether it has the expected value. Remember to add the `KernelFunction` decorator before the declaration of the function or Semantic Kernel will not recognize it:

```
    [KernelFunction, Description("Checks that each tab contains the
cells A1-A5 and B1-B5 with the correct values")]
    public static string CheckCells([Description("The file path to the
spreadsheet")] string filePath)
    {
        try
        {
            FileInfo fileInfo = new FileInfo(filePath);

            if (!fileInfo.Exists)
            {
                return "Fail: File does not exist.";
            }
            using (var package = new ExcelPackage(fileInfo))
            {
                foreach (var year in new[] { "2024", "2025" })
                {
                    var sheet = package.Workbook.Worksheets[year];
                    if (sheet.Cells["A1"].Text != "Quarter" || sheet.
Cells["B1"].Text != "Budget" ||
                        sheet.Cells["A2"].Text != "Q1" || sheet.
Cells["A3"].Text != "Q2" ||
                        sheet.Cells["A4"].Text != "Q3" || sheet.
Cells["A5"].Text != "Q4")
                    {
                        return "Fail: missing quarters";
                    }
                    for (int row = 2; row <= 5; row++)
                    {
                        if (sheet.Cells[$"B{row}"].Value is not
```

```
double)
                            {
                                  return "Fail: non-numeric values";
                            }
                      }
                }
                return "Pass";
          }
    }
    catch (Exception ex)
    {
          return $"Fail: An error occurred: {ex.Message}";
    }
}
```

The final step is to check whether the content of the numeric cells follows the two rules we defined: a budget of no more than $1M per year, and no more than a 10% increase per quarter.

Python

Again, don't forget to add the `@kernel_function` decorator so that Semantic Kernel can find your function. The function is straightforward, going from cell B2 to B5 and checking whether they increase by more than 10%, and then also checking the sum to make sure it's under $1M:

```
@kernel_function(
    description="Checks that the spreadsheet contains the correct
values, less than 1m per year and growth less than 10%",
    name="CheckValues",
)
def CheckValues(self,
              path: Annotated[str, "The path to the
spreadsheet"]) -> Annotated[str, "The result of the check"]:
    workbook = openpyxl.load_workbook(path)
    years = ['2024', '2025']
    for year in years:
        if year not in workbook.sheetnames:
            return f"Fail: Sheet for year {year} not found."
        sheet = workbook[year]
        values = [sheet[f'B{row}'].value for row in range(2, 6)]
        if not all(isinstance(value, (int, float)) for value in
values):
            return f"Fail: Non-numeric value found in sheet
{year}."
        if sum(values) >= 1000000:
            return f"Fail: Sum of values in year {year} exceeds
```

```
1,000,000."

                for i in range(len(values) - 1):
                    if values[i + 1] > values[i] * 1.10:
                        return f"Fail: More than 10% growth found from
B{i+2} to B{i+3} in sheet {year}."
            return "Pass"
```

C#

We are going to do our checks in two steps. First, we load the file and iterate through the numeric cells, adding them to an array:

```
    [KernelFunction, Description("Check that the cells B2-B5 add to
less than 1 million and don't increase over 10% each quarter")]
    public static string CheckValues([Description("The file path to
the spreadsheet")] string filePath)
    {
        try
        {
            FileInfo fileInfo = new FileInfo(filePath);
            if (!fileInfo.Exists)
            {
                return "Fail: file does not exist.";
            }
            using (var package = new ExcelPackage(fileInfo))
            {
                foreach (var year in new[] { "2024", "2025" })
                {
                    var sheet = package.Workbook.Worksheets[year];
                    if (sheet == null)
                    {
                        return "Fail: Sheet for year {year} not
found.";
                    }
                    double[] values = new double[4];
                    for (int i = 0; i < 4; i++)
                    {
                        if (sheet.Cells[i + 2, 2].Value is not double)
                        {
                            return $"Non-numeric value found in sheet
{year}.";
                        }
                        else values[i] = (double)sheet.Cells[i + 2,
2].Value;
                    }
```

Then we check whether the sum of the values of the array is over \$1M and whether there's growth of over 10%:

```
                if (sum(values) >= 1000000)
                {
                    return $"Sum of values in year {year} exceeds
1,000,000.";
                }
                for (int i = 0; i < values.Length - 1; i++)
                {
                    if (values[i + 1] > values[i] * 1.10)
                    {
                        return $"More than 10% growth found from
B{i+2} to B{i+3} in sheet {year}.";
                    }
                }
            }
            return "Pass";
        }
    }
    catch (Exception ex)
    {
        return $"An error occurred: {ex.Message}";
    }
}
```

To make the check of the sum of values simpler, create a simple helper function called sum that returns the sum of an array of doubles:

```
static double sum(double[] values)
{
    double total = 0;
    foreach (var value in values)
    {
        total += value;
    }
    return total;
}
```

By combining all the functions above in CheckSpreadsheet.py (Python) and CheckSpreadsheet.cs (C#), we can verify all the Excel files in the data/proposals folder. The entire code is in the GitHub repository for this chapter.

Besides having an Excel file, each proposal also has a Word document, of which we want to check a few aspects.

Evaluating the Word document

The proposal document will have three sections describing the team that is proposing a vaccination campaign, their experience, and the implementation details of their proposal. We want to check three aspects of the proposal document:

- Whether the team has people with experience in the medical sciences with at least one having a Ph.D. This information should be under a *Heading 1* section named Team.

- Whether the team has successfully deployed a vaccination campaign before. This information should be under a *Heading 1* section named Experience.

- Whether the project proposal has an implementation detail describing when the vaccination campaign starts and when it ends. This information should be under a *Heading 1* section named Implementation Plan.

To verify that the proposals fulfill the requirements, we are going to send the text from the Word document to a semantic function later that will tell us whether it fulfills the requirements, but first we are going to write a native function that extracts the contents of the relevant sections of each of the Word documents:

Python

To open the Python document, we need to use the python-docx package. You can install this with the following command:

```
pip install python-docx
```

Note that even though we import the docx package in the Python code, we need to install python-docx. If you install a package named docx, it will not work — that's an old, deprecated package.

We define our native function and make sure that it will be available to Semantic Kernel by decorating it with the kernel_function decorator, and its parameters and return value with the Annotated decorator:

```
from docx import Document
from typing_extensions import Annotated
from semantic_kernel.functions.kernel_function_decorator import
kernel_function

class ParseWordDocument:

    @kernel_function(
        description="Extract the text under the given heading",
        name="ExtractTextUnderHeading",
    )
```

We create a `ParseWordDocument` class to hold our native plugin, and inside that class we create a function called `ExtractTextUnderHeading`.

The remainder of the code looks for a *Heading 1* with the text `target_heading` and then reads all its paragraphs:

```python
    async def ExtractTextUnderHeading(self,
            doc_path: Annotated[str, "The path for the file we want to
evaluate"],
            target_heading: Annotated[str, "The heading we want to
extract the text from"]
            ) -> Annotated[str, "The extracted text"]:
        doc = Document(str(doc_path))
        extract = False
        extracted_text = ''

        for paragraph in doc.paragraphs:
            if paragraph.style.name == 'Heading 1':
                if extract:
                    break  # Stop if next heading is found
                extract = paragraph.text.strip().lower() ==
str(target_heading).lower()
            elif extract:
                extracted_text += paragraph.text + '\n'
        return extracted_text.strip()
```

C#

```csharp
using DocumentFormat.OpenXml.Packaging;
using DocumentFormat.OpenXml.Wordprocessing;
using System.ComponentModel;
using Microsoft.SemanticKernel;

namespace Plugins.ProposalChecker;
public class ParseWordDocument
{
    [KernelFunction, Description("Extracts the text under the heading
in the Word document")]
    public static string ExtractTextUnderHeading(string filePath,
string heading)
```

In the C# version of our plugin, we declare a `ParseWordDocument` class and a `ExtractTextUnderHeading` function. Don't forget to add the `KernelFunction` decorator, otherwise Semantic Kernel will not find your function.

The function reads all the body of the document first, and then reads all the paragraphs. Then it finds all paragraphs after a *Heading 1* that matches the `heading` parameter and returns them:

```
        using (WordprocessingDocument doc = WordprocessingDocument.
Open(filePath, false))
        {
            var body = doc.MainDocumentPart?.Document.Body;
            var paras = body?.Elements<Paragraph>();

            bool isExtracting = false;
            string extractedText = "";

            // if paras is null, return empty string
            if (paras == null)
            {
                return extractedText;
            }

            foreach (var para in paras)
            {
                if (para.ParagraphProperties != null &&
                    para.ParagraphProperties.ParagraphStyleId != null
&&
                    para.ParagraphProperties.ParagraphStyleId.Val !=
null &&
                    para.ParagraphProperties.ParagraphStyleId.Val.
Value == "Heading1" &&
                    para.InnerText.Trim().Equals(heading,
StringComparison.OrdinalIgnoreCase))
                {
                    isExtracting = true;
                    continue;
                }

                if (isExtracting)
                {
                    if (para.ParagraphProperties != null &&
                        para.ParagraphProperties.ParagraphStyleId !=
null &&
                        para.ParagraphProperties.ParagraphStyleId.Val
!= null &&
                        para.ParagraphProperties.ParagraphStyleId.Val.
Value == "Heading1")
                    {
                        break;
```

```
                }

                extractedText += para.InnerText + "\n";
            }
        }

        return extractedText.Trim();
    }
}
}
```

We will not invoke this function yet. First, we will create a semantic plugin that we will use to parse the results of the function we created above. Once we do that, we will call both functions one after another.

Developing semantic plugins

A semantic plugin is just a collection of semantic functions. Semantic functions are functions that use AI services, such as the OpenAI service or the Azure OpenAI service, to perform tasks.

> **Important – using the OpenAI services is not free**
>
> The semantic functions will call the OpenAI API. These calls require a paid subscription, and each call will incur a cost. The costs are usually small per request. GPT 3.5 costs $0.0002 per thousand tokens, but this can add up if you make a large number of calls. Note also that the prices change frequently, so make sure to check the latest prices on the following websites.
>
> OpenAI pricing: `https://openai.com/pricing`
>
> Azure OpenAI pricing: `https://azure.microsoft.com/en-us/pricing/details/cognitive-services/openai-service/`

Each plugin must have its own directory, and the name of the directory is the plugin name. The directory should contain one subdirectory per function. The name of each subdirectory is the name of the function.

Each subdirectory must contain two files: `skprompt.txt` and `config.json`. The `skprompt.txt` file contains the prompt for the semantic function that will be submitted to the AI service:

A sample skprompt.txt

The `skprompt.txt` file contains a **metaprompt**. A metaprompt is defined as a template for a prompt that will be submitted to the AI service once its variables are replaced by their values. Variables are inside double curly brackets and named with a dollar sign; for example, the `var` variable would appear as `{{$var}}` inside `skprompt.txt`.

The following example `skprompt.txt` contains three variables: `history`, `input`, and `options`:

```
[History]
{{$history}}
User: {{$input}}
------------------------------------------------
Provide the intent of the user. The intent should be one of the
following: {{$options}}
INTENT:
```

The preceding `history`, `input,` and `options` variables are just an example to show a metaprompt with multiple variables. For our grant verification use case, we will start with just one variable, called `input`, and the `skprompt.txt` file will be described in the following *Evaluating the grant proposal with a semantic plugin* section.

A sample config.json

The `config.json` file contains the parameters that control the output of the AI service.

When creating a `config.json` file, you need to specify the following parameters:

- `schema`: Reserved for future use, and should currently be set to `1`. In the future, this will be used for versioning.

- `type`: Reserved for future use, and should currently be set to `completion`. In the future, this will be used for different types of LLMs, but this has not been implemented in Semantic Kernel yet.

- `description`: An important parameter that describes what the function does. It's used by `planner` to figure out what the function can help the user accomplish.

- `execution_settings`: This is an object that contains multiple objects. You should always add an object called `"default"` inside of it. If the semantic function is called by a service with a `service_id` of `"gpt4"`, it will look for an object called `"gpt4"` inside the `execution_settings` to find which settings to use. If it doesn't find it, it will use the `execution_settings` defined under `"default"`. This allows you to configure different parameters per service.

- An OpenAI service should contain the following parameters:

  - `max_tokens`: This parameter specifies the maximum length of the response from the model. It's measured in tokens, which can be words or parts of words. For example, the word "*OpenAI*" is one token, but "*OpenAI's*" might be two tokens ("*OpenAI*" and "*'s*"). Setting a higher number of `max_tokens` allows for longer responses, but also consumes more computational resources. A good rule of thumb is to assume that tokens add 33% to the number of words, so a text with 75 words will need approximately 100 tokens.

- `temperature`: The temperature controls the randomness of the model's output. A lower temperature (close to 0) makes the model's responses more predictable and deterministic, while a higher temperature (closer to 1) increases randomness and creativity in the responses. A temperature of 0 means the model will always choose the most likely next token, whereas at 1, it chooses more freely.

- `top_p` (also known as nucleus sampling): Helps in controlling the diversity of the generated text. It works by only considering the most likely next tokens that cumulatively make up the probability *p*. For instance, if `top_p` is set to 0.9, the model will choose the next token from the top 90% of the probability distribution. Lower values of `top_p` make the model more conservative in its choices, while higher values allow for more diversity and less predictability.

- `presence_penalty`: Encourages the model to mention new topics and introduce variety. A higher `presence_penalty` value will make the model less likely to repeat the same topics or entities that have already been discussed, leading to a wider range of subjects in the output.

- `frequency_penalty`: Similar to `presence_penalty`, `frequency_penalty` discourages the model from repeating the same lines or information. A higher `frequency_penalty` value makes the model less likely to use the same words or phrases frequently, promoting more varied and less repetitive language.

Differences between presence penalty and frequency penalty

Although the `presence_penalty` and `frequency_penalty` parameters look similar, they work in slightly different ways. The `presence_penalty` parameter is about topics, and when it's high, it promotes topic diversity. For example, if the answer is about technology and it contains the topic *"artificial intelligence,"* a high `presence_penalty` value would try to prevent that topic from appearing again. The `frequency_penalty` is about words and linguistic variety. For example, if the answer already contains *"cutting-edge technology,"* a high `frequency_penalty` value would encourage the model to find other words to express the same idea, such as *"new techniques,"* instead of repeating *"cutting-edge technology"* again.

- `input_variables`: This attribute contains a single array called `parameters`. Each parameter has three components:

  - `name`: The name of the parameter. For the following `config.json` file, the names would be `input`, `history`, and `options`.

  - `description`: An important parameter that describes what the parameter is. It's used by the planner to figure out which functions to use and which parameters to pass to them.

  - `required`: This specifies whether the value is required or not.

The following is a completed example of a `config.json` file:

```json
{
    "schema": 1,
    "type": "completion",
    "description": "Gets the intent of the user.",
    "execution_settings": {
        "default": {
            "temperature": 0.8,
            "number_of_responses": 1,
            "top_p": 1,
            "max_tokens": 4000,
            "presence_penalty": 0.0,
            "frequency_penalty": 0.0
        },
        "gpt4": {
            "temperature": 0.8,
            "number_of_responses": 1,
            "top_p": 1,
            "max_tokens": 4000,
            "presence_penalty": 0.0,
            "frequency_penalty": 0.0
        }
    },
    "input_variables":[
            {
                "name": "input",
                "description": "The user's request.",
                "required": true
            },
            {
                "name": "history",
                "description": "The user's previous requests.",
                "required": true
            },
            {
                "name": "options",
                "description": "Any options the user has already
told us about.",
                "required": false
            }

]
```

```
        }
    }
```

Now that we know the details of creating `config.json` files, let's create a semantic plugin to evaluate proposals.

Evaluating the grant proposal with a semantic plugin

Our semantic plugin will use the following directory structure:

```
└──ch3
   └──plugins
      └──ProposalChecker
          ├──CheckQualifications
          │       ├──skprompt.txt
          │       └──config.json
          ├──CheckExperience
          │       ├──skprompt.txt
          │       └──config.json
          ├──CheckImplementationDescription
          │       ├──skprompt.txt
          │       └──config.json
```

As explained at the beginning of the chapter, the three semantic functions we will implement are as follows:

- `CheckQualifications`: Determines whether the team requesting the grant has the required academic qualifications to perform the vaccination campaign
- `CheckExperience`: Determines whether the team requesting the grant has performed a successful campaign before
- `CheckImplementationDescription`: Determines whether the proposal contains reasonable deadlines by which to implement the campaign

The following are the `config.json` and the `skprompt.txt` files for all these functions. For simplicity, we will use the same configuration for the `config.json` file for all three functions.

Implementing the semantic functions

We're going to implement a semantic function that checks whether the project team fulfills the required qualifications for the project. The qualifications are that one of the members has to have training in the medical sciences, and at least one other should have a Ph.D.

First, we have the `config.json` file for the first semantic function, `CheckQualifications`:

config.json for CheckQualifications

```json
{
    "schema": 1,
    "type": "completion",
    "description": "Check whether project managers fulfill the
required qualifications",
    "execution_settings": {
        "default": {
            "temperature": 0.8,
            "number_of_responses": 1,
            "top_p": 1,
            "max_tokens": 4000,
            "presence_penalty": 0.0,
            "frequency_penalty": 0.0
        }
    },
    "input_variables": [
        {
            "name": "input",
            "description": "Biographies of the project managers
including their qualifications",
            "defaultValue": ""
        }
    ]
}
```

skprompt.txt for CheckQualifications

The metaprompt receives the biographies of the project team as an input and checks whether they have the required qualifications, returning `"No"` if they don't and `"Yes"` if they do:

```
These are the biographies of the project team:
{{$input}}


We require at least one of the project team members to have a Ph.D.
and at least one to have experience in the medical sciences.

If the project team does not meet these requirements, say "No".
If the project team meets these requirements, say "Yes"
```

Next, we will show the `config.json` file for the second semantic function, CheckExperience:

config.json for CheckExperience

We're just going to change the description of the function and the parameters. To save space, we show just the changes in the following code. All remaining attributes are the same as the ones for CheckQualifications. You can check the GitHub repository for the complete file:

```
"description": "Check whether the project team has the required
experience deploying a project of this size and complexity",

    "input_variables": [
        {
            "name": "input",
            "description": "Description of an earlier project",
            "defaultValue": ""
        }
    ]
```

skprompt.txt for CheckExperience

The metaprompt receives the experience of the project team as an input and checks whether they have the required experience, returning "No" if they don't and "Yes" if they do:

```
The project team provided this experience:
{{$input}}

Do they have enough experience to conduct a massive vaccination
campaign in a new country?

If they have had a successful experience in Atlantis or another large
country, respond "Yes", otherwise respond "No".
```

Finally, we show the `config.json` file for the second semantic function, CheckImplementationDescription.

config.json for CheckImplementationDescription

For the configuration of this function, we are just going to change the description of the function and the parameters. To save space, we show just the changes in the following code. All remaining attributes are the same as the ones for CheckQualifications.

You can check the GitHub repository for the complete file:

```
"description": "Check whether the project description includes
reasonable dates in their project implementation",
    "input_variables": [
        {
            "name": "input",
            "description": "Description of the implementation of the
project",
            "required": true
        }
    ]
```

skprompt.txt for CheckImplementationDescription

The metaprompt for `CheckImplementationDescription` checks whether the project has reasonable dates between 2024 and 2025:

```
The project team provided this implementation description:
{{$input}}

Can they fulfill the vaccination campaign in two years, starting in
2024 and ending in 2025?

If there are no dates listed, say No.
If the dates are outside of the 2024-2025 range, say No.
If there are dates between 2024 and 2025, say Yes.
```

Loading and testing the semantic functions

Now that we have created both a native function that extracts text from a Word document and a set of semantic functions to analyze its contents, we just need to call them in sequence to check whether the proposal fulfills the requirements.

C#

We start by creating a helper function that given a Kernel, a document, a header, and a function name, returns the result of checking the header with the function:

```
async Task<string> CheckDocumentPart(Kernel kernel, string path,
string part, string function)
{
    KernelPlugin documentPlugin = kernel.Plugins["ParseWordDocument"];
    KernelFunction documentParser =
documentPlugin["ExtractTextUnderHeading"];
```

```
    KernelPlugin documentReader = kernel.Plugins["ProposalChecker"];

    var contextVariables = new KernelArguments
    {
        ["filePath"] = path,
        ["heading"] = part
    };
    // Check for text
    var text = await kernel.InvokeAsync(documentParser,
contextVariables);

    var contextVariables2 = new KernelArguments
    {
        ["input"] = text.ToString(),
    };

    var result = await kernel.InvokeAsync(documentReader[function],
contextVariables2);
    return result.ToString();
}
```

We then call that function with each of the example documents we want to test. For brevity, the following code includes just one call, but you can see the tests for all documents in the GitHub repository:

```
var docPath1 = $"{data_directory}/correct/correct.docx";

string result_experience = CheckDocumentPart(kernel, docPath1,
"Experience", "CheckExperience").Result;
string result_qualifications = CheckDocumentPart(kernel, docPath1,
"Team", "CheckQualifications").Result;
string result_implementation = CheckDocumentPart(kernel, docPath1,
"Implementation", "CheckImplementationDescription").Result;

Console.WriteLine($"Checking {docPath1}");
Console.WriteLine($"Experience: {result_experience}");
Console.WriteLine($"Qualifications: {result_qualifications}");
Console.WriteLine($"Implementation: {result_implementation}");
```

The results are as follows:

```
Experience: Yes
Qualifications: Yes
Implementation: Yes
```

The following code shows how to run both the Excel checks and the Word checks in Python:

Python

In Python, we import the `semantic_kernel` libraries and the modules that contain our native functions: `CheckSpreadsheet` and `ParseWordDocument`:

```
import asyncio
from semantic_kernel.connectors.ai.open_ai import OpenAIChatCompletion
import semantic_kernel as sk
from semantic_kernel.utils.settings import openai_settings_from_dot_
env
from CheckSpreadsheet import CheckSpreadsheet
from ParseWordDocument import ParseWordDocument
from semantic_kernel.functions.kernel_arguments import KernelArguments
```

To make things simpler, we create two helper functions. The first helper function, `run_spreadsheet_check`, will check whether a spreadsheet is correct. To do so, it imports the `CheckSpreadsheet` native plugin, and given a `path` variable representing the path of a spreadsheet file and the name of a native `function`, runs the native function on the file that `path` points to:

```
async def run_spreadsheet_check(path, function):
    kernel = sk.Kernel()

    check_spreadsheet = kernel.add_plugin(CheckSpreadsheet(),
"CheckSpreadsheet")

    result = await kernel.invoke(
        check_spreadsheet[function], KernelArguments(path = path)
    )
    print(result)
```

The second helper function, `run_document_check`, is used to check Word documents. It creates a semantic kernel and adds the GPT-3.5 service to it. Then, it attaches the `ParseWordDocument` native plugin and the Proposal Checker semantic plugin.

The function receives four parameters:

- `path`: The path of the document we want to check
- `function`: The native function that we are going to use to parse the document
- `target_heading`: The heading we want to extract from the document
- `semantic_function`: The semantic function that we want to use to evaluate the text extracted from the document

- Later, we will run this function once per header we want to parse. For example, if we call the function with the following parameters, we would open the correct.docx document, use the ExtractTextUnderHeading native function to extract the text under the heading "Experience", and then run the CheckExperience semantic function on the extracted text:

```
run_document_check("correct.docx", "ExtractTextUnderHeading",
"Experience", "CheckExperience")
```

Let's see how to define a function that does all that. We first create the kernel and add the plugins:

```
async def run_document_check(path, function, target_heading, semantic_
function):
    kernel = sk.Kernel()
    api_key, org_id = openai_settings_from_dot_env()
    gpt35 = OpenAIChatCompletion("gpt-3.5-turbo", api_key, org_id,
service_id = "gpt35")
    kernel.add_service(gpt35)

    parse_word_document = kernel.add_plugin(ParseWordDocument(),
"ParseWordDocument")

    text = await kernel.invoke(
        parse_word_document[function],
        KernelArguments(doc_path = path, target_heading = target_
heading)
    )

    check_docs = kernel.add_plugin(None, "ProposalChecker", "../../
plugins")
    result = await kernel.invoke(check_docs[semantic_function],
KernelArguments(input = text))
    print(f"{target_heading}: {result}")
```

Still inside the run_document_check helper function, we use the native function we defined in the *Evaluating the Word document* section of this chapter to extract the whole text of the document:

```
text = await kernel.run_async(
    parse_word_document[function],
    input_vars=variables
)
```

Now that we have the text, we run the function passed as a parameter of run_document_check to check it has the contents we want:

```
check_docs = kernel.import_semantic_skill_from_directory("../
plugins", "ProposalChecker")
    result = check_docs[semantic_function](str(text))
    print(f"{target_heading}: {result}")
```

Now back to the main program, the following code uses the two helper functions defined previously to check several documents. For brevity, we show just a few calls here, but the source code in the GitHub repository does all the checks:

```python
async def main():
    data_path = "../data/proposals/"
    await run_spreadsheet_check(f"{data_path}/correct/correct.xlsx",
"CheckTabs")

    await run_spreadsheet_check(f"{data_path}/incorrect2/over_budget.
xlsx", "CheckCells")

    await run_spreadsheet_check(f"{data_path}/incorrect3/fast_
increase.xlsx", "CheckValues")

    print("Word document checks:")

    await run_document_check(f"{data_path}/correct/correct.docx",
"ExtractTextUnderHeading", "Experience", "CheckExperience")
    await run_document_check(f"{data_path}/correct/correct.docx",
"ExtractTextUnderHeading", "Team", "CheckQualifications")
    await run_document_check(f"{data_path}/correct/correct.
docx", "ExtractTextUnderHeading", "Implementation",
"CheckImplementationDescription")


# Run the main function
if __name__ == "__main__":
    asyncio.run(main())
```

The results are as follows:

```
Experience: Yes
Qualifications: Yes
Implementation: Yes
```

Summary

In this chapter, you learned how to create native functions and combine them into plugins. You also learned how to create semantic functions and all the details of their parameters, and how to combine them into plugins.

In the next chapter, we will learn how to generate images from Semantic Kernel, and we will also learn how to use the functions covered here with slight modifications to create a pipeline that can process many documents.

4

Performing Complex Actions by Chaining Functions

In the previous chapter, we learned how to create native plugins, which we used to check the format of Excel and Word documents, and semantic plugins, which we used to verify whether the content of the documents fulfilled our requirements.

In this chapter, we will start by creating a simple pipeline that generates images. The pipeline will receive a text with clues about an animal and will then generate a text that guesses the animal from the clues, as well as generating a picture of the animal.

Later in the chapter, we will continue the application of the previous chapter: verifying whether grant requests fulfill some requirements. For that application, a grant request will come with two files inside a folder: a Word document and an Excel spreadsheet.

Our application checks that the Word document contains a proposal for a vaccination campaign, including the team who will perform it and their experience, and an Excel file that contains its budget.

This type of scenario is common in enterprises: governments and corporations must prioritize requests for projects, editors must approve or reject book proposals, and lawyers must verify that the documents they are filing in court fulfill legal requirements. It's not uncommon to have to verify thousands or tens of thousands of documents. This is a kind of job that, until recently, was done manually. We're going to write a pipeline that automates it.

In this chapter, we'll be covering the following topics:

- Creating a native plugin that generates images
- Chaining a semantic plugin that outputs text with the native plugin that generates images
- Running a complex, multistep pipeline

By the end of the chapter, you will have the tools required to build a kernel that can perform many coordinated functions, and that can be used as a copilot or be integrated into other tools, such as a planner.

Technical requirements

To complete this chapter, you will need to have a recent, supported version of your preferred Python or C# development environment:

- For Python, the minimum supported version is Python 3.10, and the recommended version is Python 3.11
- For C#, the minimum supported version is .NET 8

In this chapter, we will call OpenAI services. Given the amount that companies spend on training these LLMs, it's no surprise that using these services is not free. You will need an **OpenAI API** key, either directly through **OpenAI** or **Microsoft**, via the **Azure OpenAI** service.

If you are using .NET, the code for this chapter is at `https://github.com/PacktPublishing/ Building-AI-Applications-with-Microsoft-Semantic-Kernel/tree/main/ dotnet/ch4`.

If you are using Python, the code for this chapter is at `https://github.com/PacktPublishing/ Building-AI-Applications-with-Microsoft-Semantic-Kernel/tree/main/ python/ch4`.

You can install the required packages by going to the GitHub repository and using the following: `pip install -r requirements.txt`.

Creating a native plugin that generates images

To learn about the power of chaining functions, we are going to create functions that perform very different actions. We will start by creating functions that generate images and putting them in a plugin. Then, we're going to learn how to incorporate these functions into a more complex chain.

In some applications, you may want to generate an image with AI. For example, social media posts with images tend to get more engagement, but creating images without AI or finding images can be time-consuming and expensive.

Compared to market prices of non-AI images, generating images with AI is very cheap. On the other hand, generating images is still one of the most compute-intensive activities that can be done with AI. Recent research from Hugging Face [1] has shown that generating an image is 2,000 times more expensive in terms of carbon emissions than generating a text answer. These costs will be passed down to you.

> **Costs of OpenAI image generation**
>
> If you want to reproduce the content in this section, be aware that image generation is far more costly than text generation. You will need an API key, and each image generation costs $0.04 per image for the following examples, and up to $0.12 per image if you want to create higher-quality images with higher resolutions.
>
> Prices change frequently, and you can check the latest prices at `https://openai.com/pricing`.

In *Figure 4.1*, we show two examples of images generated with AI that I have created to enhance a couple of my social media posts. The first, me as a cowboy, was for a post about billionaires using cowboy hats. The second, with me as a character on the cover of a romantic novel, was for a post about writing fiction. Each image took less than a minute to generate with AI. I ran a test in Threads, Instagram's new microblogging app where I have over 10,000 followers, and the image posts had multiple times more engagement than posts with the same text but no image.

Figure 4.1 – Images generated with AI for social media posts

Image generation support in Microsoft Semantic Kernel is not consistent. The C# API has an object called `TextToImage` that can generate images using DALL-E 2, a model released in November 2022, but that API is not available for Python. That image-generating model is now obsolete, having been superseded by DALL-E 3 in October of 2023; however, at the time of writing, Semantic Kernel does not offer an out-of-the-box way to access DALL-E 3.

We will create two native plugins, one for C# and one for Python, that allow us to access DALL-E 3 from Semantic Kernel.

> **Adding new models to the kernel**
>
> Although we are using DALL-E 3, as an interesting and novel model, as the example that we are adding to the kernel, this approach of creating a native plugin wrapper works for any model that has an API, including Claude from Anthropic, Gemini from Google, and hundreds of models from Hugging Face. Any AI service that is made available through a REST API can be added in this way.

Writing a DALL-E 3 wrapper in Python

It is easier to write a DALL-E 3 wrapper in Python than in C# because OpenAI offers and supports a Python package called openai that allows developers to access any new OpenAI function as soon as it's made available. All we must do is create a native function that uses the OpenAI package and send a request to DALL-E 3:

```python
from dotenv import load_dotenv
from openai import OpenAI
import os
from semantic_kernel.skill_definition import kernel_function

class Dalle3:
    @kernel_function(
        description="Generates an with DALL-E 3 model based on a
prompt",
        name="ImageFromPrompt",
        input_description="The prompt used to generate the image",
    )
    def ImageFromPrompt(self, input: str) -> str:
        load_dotenv()
        client = OpenAI(api_key=os.getenv("OPENAI_API_KEY"))
```

As we saw in *Chapter 3*, we declare a native function using the `kernel_function` decorator. We then simply instantiate an OpenAI client object with our API key.

Now let's submit the request:

```python
        response = client.images.generate(
            model="dall-e-3",
            prompt=input,
            size="1024x1024",
            quality="standard",
            n=1,
        )

        image_url = response.data[0].url
        return image_url
```

The Python OpenAI client contains the `images.generate` method, which will call DALL-E 3 and return the generated URL. We simply call it and return the URL.

Now, we are going to create a simple script that instantiates the plugin and calls it:

```python
import asyncio
import semantic_kernel as sk
from OpenAiPlugins import Dalle3

async def main():
    kernel = sk.Kernel()
    animal_str = "A painting of a cat sitting in a sofa in the
impressionist style"
    dalle3 = kernel.import_skill(Dalle3())

    animal_pic_url = await kernel.run_async(
        dalle3['ImageFromPrompt'],
        input_str=animal_str
    )

    print(animal_pic_url)

if __name__ == "__main__":
    asyncio.run(main())
```

The preceding code instantiates the `Dalle3` native plugin and calls its `ImageFromPrompt` function with the `"A painting of a cat sitting on a sofa in the impressionist style"` input parameter. An example output is in *Figure 4.2*:

Figure 4.2 – A cat in the impressionist style generated by DALL-E 3 by Python

Now that we have seen how to do this in Python, let's see how to do it in C#.

Writing a DALL-E 3 wrapper in C#

OpenAI does not provide a supported package for C# that allows users of that language to interact with its service. The best way of interacting with the OpenAI service for C# users is to use Microsoft Semantic Kernel, which has a `TextToImage` functionality, but at the time of writing, it only provides connectivity to DALL-E 2.

Most AI services will expose a REST API. Therefore, to connect Microsoft Semantic Kernel to them, one solution is to write a native plugin that wraps the REST API. We show how to do this for DALL-E 3 here:

```
using System.ComponentModel;
using System.Net.Http.Headers;
using System.Net.Http.Json;
using Microsoft.SemanticKernel;
using System.Text;
using System.Text.Json;
using System.Text.Json.Nodes;


namespace Plugins;


public class Dalle3
{
    [KernelFunction, Description("Generate an image from a prompt")]
    async public Task<string> ImageFromPrompt([Description("Prompt
describing the image you want to generate")] string prompt)
    {
```

Making a REST POST request requires several packages, such as `System.Text.Json` and `System.Net`. Like what we did in *Chapter 3*, we use a decorator, `KernelFunction`, to signal that the function is accessible to Semantic Kernel, and a `Description` attribute to describe what our function does.

We then create an `HttpClient` object. This object will make a REST API call. We need to set it up with our API key as a `Bearer` token and set its header as accepting "`application/json`" because that's how the OpenAI API will respond:

```
        HttpClient client = new HttpClient
        {
            BaseAddress = new Uri("https://api.openai.com/v1/")
        };

        var (apiKey, orgId) = Settings.LoadFromFile();

        client.DefaultRequestHeaders
```

```
            .Accept
            .Add(new MediaTypeWithQualityHeaderValue("application/
json"));

        client.DefaultRequestHeaders.Authorization = new
AuthenticationHeaderValue("Bearer", apiKey);
        client.DefaultRequestHeaders.Accept.Add(new
MediaTypeWithQualityHeaderValue("application/json"));
```

The next step is to submit the POST request to the API:

```
        var obj = new {
            model = "dall-e-3",
            prompt = prompt,
            n = 1,
            size = "1024x1024"};

        var content = new StringContent(JsonSerializer.Serialize(obj),
Encoding.UTF8, "application/json");

        var response  = await client.PostAsync("images/generations",
content);
```

We created a JSON object, obj, using the fields that are required by the OpenAI API. The model field states what model we're using, and here we make sure to specify "dall-e-3" to use DALL-E 3. The documentation of all the possible parameters can be found here: https://platform. openai.com/docs/api-reference/images/create.

The final step is to recover the url field from the JSON returned by OpenAI. That url field points to the image:

```
        if (!response.IsSuccessStatusCode)
        {
            return $"Error: {response.StatusCode}";
        }


        string jsonString = await response.Content.
ReadAsStringAsync();
        using JsonDocument doc = JsonDocument.Parse(jsonString);
        JsonElement root = doc.RootElement;
        return root.GetProperty("data")[0]!.GetProperty("url")!.
GetString()!;
    }
}
```

Next, let's see how to call the plugin:

```
using Microsoft.SemanticKernel;
using Plugins;

var (apiKey, orgId) = Settings.LoadFromFile();

var builder = Kernel.CreateBuilder();
builder.Plugins.AddFromType<Dalle3>();
var kernel = builder.Build();

string prompt = "A cat sitting on a couch in the style of Monet";
string? url = await kernel.InvokeAsync<string>(
    "Dalle3", "ImageFromPrompt", new() {{ "prompt", prompt }}
);

Console.Write(url);
```

To call the plugin, we added a reference to the source file, `Plugins`, instantiated the `Dalle3` plugin with `AddFromType`, and called its `ImageFromPrompt` method, passing `prompt` as a parameter.

The resulting picture is the following:

Figure 4.3 – A cat in the impressionist style generated by the C# native plugin

Now that we have created a function that accesses a new service, let's incorporate it into a solution that uses it.

Using multiple steps to solve a problem

Although programming solutions step by step can be very helpful, one of the best abilities that Semantic Kernel gives users is allowing them to make requests using natural language. This will require using **planners**, which we will use in *Chapter 5*, to break down a user request into multiple steps and then automatically call each step in the appropriate order.

In this section, we will solve problems by telling Semantic Kernel which functions to call. This is helpful for making sure that the solutions we make available to the planner work, and it is also helpful when we want to explicitly control how things are executed.

To illustrate the manual approach, we will see how to give Semantic Kernel clues about an animal, guess it with a semantic function, and then generate an image of the animal using the native function we created in the previous section.

Generating an image from a clue

In the following code, we have two steps. In the first step, we will use GPT-3.5 to guess an animal from clues. To do that, we will create a semantic plugin called `AnimalGuesser`.

> **Important: Using OpenAI services is not free**
>
> The semantic functions will call the OpenAI API. These calls require a paid subscription, and each call will incur a cost. The costs are usually small per request. GPT 3.5 costs $0.0002 per thousand tokens, but they may add up if you make a large number of calls. Prices change frequently, so make sure to check the latest prices on the following websites:
>
> OpenAI pricing: `https://openai.com/pricing`
>
> Azure OpenAI pricing: `https://azure.microsoft.com/en-us/pricing/details/cognitive-services/openai-service/`

The semantic plugin, as always, consists of two files, `config.json` and `skprompt.txt`, listed as follows:

config.json

```json
{
    "schema": 1,
    "name": "GuessAnimal",
    "type": "completion",
    "description": "Given a text with clues, guess the animal",
    "execution_settings": {
        "default": {
```

```
            "temperature": 0.8,
            "number_of_responses": 1,
            "top_p": 1,
            "max_tokens": 4000,
            "presence_penalty": 0.0,
            "frequency_penalty": 0.0
        }
    },
    "input_variables": [
        {
            "name": "input",
            "description": "CLues about an animal",
            "required": true
        }
    ]
}
```

skprompt.txt

```
Below, there's a list of clues about an animal.

{{$input}}

From the clues above, guess what animal it is.

Provide your answer in a single line, containing just the name of the
animal.
```

As we always do, we must make sure that the `description` fields in the `config.json` files are set correctly. This will not have any effect now, but when we start using the planner or letting Semantic Kernel automatically call functions, the kernel will use the `description` fields to figure out what each function does and decide which ones to call.

For now, let's see how to tell the kernel to call functions in sequence.

Chaining semantic and native functions with C#

In C#, you need to use the `KernelFunctionCombinators` class to create a function pipeline. The code for the class is provided in the GitHub repository.

The code for implementing a function pipeline follows:

```
using Microsoft.SemanticKernel;
using Plugins;

var (apiKey, orgId) = Settings.LoadFromFile();

var builder = Kernel.CreateBuilder();
builder.Plugins.AddFromType<Dalle3>();
builder.AddOpenAIChatCompletion("gpt-3.5-turbo", apiKey, orgId);
var kernel = builder.Build();

KernelPlugin animalGuesser = kernel.
ImportPluginFromPromptDirectory("../../../plugins/AnimalGuesser");
string clues = "It's a mammal. It's a pet. It meows. It purrs.";
```

In the preceding snippet, we create our kernel, add an OpenAI service to it, and add the AnimalGuesser and Dalle3 plugins to it.

Next, we assign the functions we want to call, AnimalGuesser.GuessAnimal and Dalle3.ImageFromPrompt, to KernelFunction variables:

```
KernelFunction guessAnimal = animalGuesser["GuessAnimal"];
KernelFunction generateImage = kernel.Plugins["Dalle3"]
["ImageFromPrompt"];

KernelFunction pipeline = KernelFunctionCombinators.Pipe(new[] {
    guessAnimal,
    generateImage
}, "pipeline");
```

Lastly, we create a KernelArguments object called context and pass it as a parameter to InvokeAsync:

```
KernelArguments context = new() { { "input", clues } };
Console.WriteLine(await pipeline.InvokeAsync(kernel, context));
```

The attributes of the context object must match what the first function is expecting. In our case, the AnimalGuesser.GuessAnimal function expects a parameter named input. From then on, the pipeline will call each function, get the output as a text string, and pass that text string as the first parameter to the next function. In our case, even though the first parameter of the Dalle3.ImageFromPrompt function is called prompt instead of input, the call is still going to work. You only need to provide the correct name for the parameter used in the first step of the pipeline.

If you run the preceding program, you will get a picture of a cat:

Figure 4.4 – Picture of a cat generated from AI guessing the animal from clues

Chaining semantic and native functions with Python

As we did in C#, let's use Python to create a script that starts from a list of clues, guesses the animal that the clues refer to, and then generates a picture of the animal.

We will build on the plugins we already have. We will reuse the native plugin that we created to generate images using DALL-E 3.

One new thing is that we are going to create a function called `pipeline` that receives a list of functions and an input parameter and then calls each function in the list, passing the output of the call as the input parameter of the next function in the list.

The definition of the function is as follows:

```
async def pipeline(kernel, function_list, input):
    for function in function_list:
        args = KernelArguments(input=input)
        input = await kernel.invoke(function, args)
    return input
```

We start as we always do, creating a kernel and adding an AI service to it. Here, we are assigning the plugins to variables, which will enable us to reference functions in the next step:

```
import asyncio
from semantic_kernel.connectors.ai.open_ai import OpenAIChatCompletion
import semantic_kernel as sk
from OpenAiPlugins import Dalle3

async def main():
    kernel = sk.Kernel()
    api_key, org_id = sk.openai_settings_from_dot_env()
    gpt35 = OpenAIChatCompletion("gpt-3.5-turbo", api_key, org_id)
    kernel.add_chat_service("gpt35", gpt35)

    generate_image_plugin = kernel.import_skill(Dalle3())
    animal_guesser = kernel.import_semantic_skill_from_
directory("../../plugins", "AnimalGuesser")
```

With that, we can now give our model the clues and ask it to guess:

```
    clues = """
    I am thinking of an animal.
    It is a mammal.
    It is a pet.
    It is a carnivore.
    It purrs."""

    function_list = [
        animal_guesser['GuessAnimal'],
        generate_image_plugin['ImageFromPrompt']
    ]

    animal_pic_url = await pipeline(kernel, function_list, clues)
    print(animal_pic_url)

if __name__ == "__main__":
    asyncio.run(main())
```

As seen in the preceding snippet, to run the pipeline, we create the list of functions we want to call in order, adding it to the `function_list` variable, and then call the `pipeline` function. The `pipeline` function will run the first function with the input parameter you passed, then it will use the output of the first function as the parameter of the second function, and so on.

Like what happened in the C# example, the output will be a URL pointing to a freshly generated picture of a cat (not displayed).

Now that we're done with a simple example of a pipeline, let's go back to the problem we were solving in *Chapter 3* and described in the introduction of this chapter: verifying whether a proposal for a vaccination campaign fulfills the basic requirements.

Dealing with larger, more complex chains

In the previous chapter, we created three plugins:

- `CheckSpreadsheet`: A native plugin that checks that the Excel spreadsheet contains the required fields and that they fulfill some rules
- `ParseWordDocument`: A native plugin that extracts text from a Word document
- `ProposalChecker`: A semantic plugin that checks whether text blocks fulfill some requirements, such as "*does this text block describe a team that has a Ph.D. and a medical doctor?*"

With these three plugins, you can already solve the business problem of checking proposals by calling each plugin separately and writing the logic to handle whether there was an error. This is likely sufficient for problems that have a small number of steps.

While we are still going to use a small number of steps and a small number of documents for didactic purposes, the approach to analyzing and making decisions on a large number of documents presented in this chapter excels when there are many steps and many documents to process.

Let's see how to implement it.

Preparing our directory structure

Before we start, we need to make sure that we have the data for the proposals loaded in the `data/ proposals` folder. We will also reuse the native plugins by putting them in the same directory as our main program. The semantic plugins will be in the `plugins` directory.

We will modify our native and semantic functions slightly from what we did in the previous chapter. The main change is that we will introduce error handling directly into the semantic functions, which will enable us to process many documents with a single call.

Below is the directory structure of our solution. Each proposal is represented by a directory in the `proposals` directory, and each directory should contain exactly two files, one Excel file with the extension `.xlsx` and one Word file with the extension `.docx`:

```
└──data
    └──proposals
    │   │──correct
```

```
|    |        ├────correct.docx
|    |        └────correct.xlsx
|    ├────incorrect01
|    |        ├────missing_experience.docx
|    |        └────incorrect_template.xlsx
|    ├────incorrect02
|    |        ├────missing_qualifications.docx
|    |        └────over_budget.xlsx
|    └────(...)
|    ├──── incorrect10
└────plugins
         └────ProposalCheckerV2
                  ├────CheckDatesV2
                  |        ├────skprompt.txt
                  |        └────config.json
                  ├────CheckPreviousProjectV2
                  |        ├────skprompt.txt
                  |        └────config.json
                  └────CheckTeamV2
                           ├────skprompt.txt
                           └────config.json
└────ch4
     └────code
              ├────python
              |        ├────ch4.py
              |        ├────ParseWordDocument.py
              |        └────CheckSpreadsheet.py
              └────dotnet
                       ├────Program.cs
                       ├────ParseWordDocument.cs
                       └────CheckSpreadsheet.cs
```

This follows the same structure we used earlier: a main file containing the code (ch4.py or Program.cs), additional files in the same directory, each containing a native plugin (ParseWordDocument and CheckSpreadsheet), and all the semantic plugins in a dedicated directory, plugins. We separate folders by language because that makes it simpler to manage the virtual environments that hold installed packages by folder. Semantic plugins are language-independent and can have their own directory.

Now that we have described the expected directory structure, let's look at the high-level flow of our process.

Understanding the flow of our process

We will start by writing a native plugin called `Helpers` that contains a native function called `ProcessProposalFolder`, which when given a path that represents a folder, checks whether it contains exactly one Excel file and a Word document. If it does, it returns the path of the folder, if not, it returns a string with an error.

Once we create the `Helpers` plugin, we will be almost ready to call the functions we developed in *Chapter 3*. We will make two modifications to the existing files `ParseWordDocument` and `CheckSpreadsheet`.

One modification we will make to both files will be to check if the input is an error state. If it is, we simply pass the error state forward. If we are not in an error state, we keep passing the folder path forward. We will need to make these simple modifications to all the native functions and semantic functions.

The second and last modification will be to the `ParseWordDocument` native plugin. We will add three separate helper functions, each one parsing one of the three different required sections of the document (`Team`, `Experience`, and `Implementation` details). The new functions will simply call the existing function with a parameter representing one section per function.

The reason for doing all of this is to only have functions with a single parameter in the pipeline. This enables the return of each function to be passed as a parameter to the next function, which will make things much simpler.

The full pipeline, with 10 steps, is represented in the next diagram.

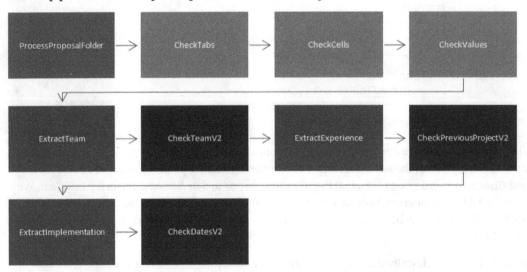

Figure 4.5 – Visual representation of the pipeline

In *Figure 4.5*, orange boxes represent native functions that deal with the filesystem, green boxes represent native functions that deal with Excel, blue boxes represent native functions that deal with Word, and purple boxes represent semantic functions.

Now that we understand all the steps for the pipeline, let's create the new plugin and function and make the required modifications.

Creating the native function to process a folder

To get the pipeline started, we need a plugin that ensures that the folder that we want to process contains the correct files. Since this is an activity that requires interacting with the operating system, we need to create a native function, which we will host inside a plugin we will call `Helpers`.

The code for the plugin is as follows:

C#

```
using Microsoft.SemanticKernel;
using System.ComponentModel;
using System.ComponentModel.DataAnnotations;

public class Helpers
{
    [KernelFunction, Description("Checks that the folder contains one
Word and one Excel file")]
    public static string ProcessProposalFolder([Description("Folder
potentially containing")] string folderPath)
    {
```

As usual, since we want this function to be available to Semantic Kernel, we use the `KernelFunction` decorator and the descruiption of what the function does under the `Description` variable.

Now, we are simply going to count the number of files available in the folder. Remember that we want exactly one file with the `.docx` extension and one file with the `.xlsx` extension:

```
        string result = folderPath;

        if (!Directory.Exists(folderPath))
        {
            return "Error: Folder does not exist";
        }

        var files = Directory.GetFiles(folderPath);
        int wordCount = files.Count(f => Path.GetExtension(f).
ToLower() == ".docx");
        int excelCount = files.Count(f => Path.GetExtension(f).
ToLower() == ".xlsx");

        if (wordCount == 1 && excelCount == 1)
```

```
        {
            return result;
        }
        else if (wordCount == 0 && excelCount == 0)
        {
            return "Error: Folder does not contain one Word and one
Excel file";
        }
        else if (wordCount == 0)
        {
            return "Error: Folder missing Word file";
        }
        else if (excelCount == 0)
        {
            return "Error: Folder missing Excel file";
        }
        return "Error: Folder contains more than one Word or Excel
file";

    }
}
```

The logical part of the function is very simple. It counts the number of files with the `.docx` and `.xlsx` extensions in the directory. If the directory has one of each, the call succeeded. We signal success by passing the folder as the return result. In any other situation, we generate a string with an error message. We will use the convention that error messages in this pipeline start with `Error`.

Let's now check out the code for the plugin in Python:

Python

```python
from typing_extensions import Annotated
from semantic_kernel.functions.kernel_function_decorator import
kernel_function
import os

class Helpers:

    @kernel_function(
        description="Checks that the folder contains the expected
files, an Excel spreadsheet and a Word document",
        name="ProcessProposalFolder"
    )
    def ProcessProposalFolder(self, input: Annotated[str, "The file
path to the folder containing the proposal files"]) -> str:
```

In Python, we use the `kernel_function` decorator to indicate that this function can be used by Semantic Kernel. We also add an `Annotated` description to the parameter.

Then, the function code is very simple. Similar to the C# function above, we count how many files with the `.docx` and `.xlsx` extensions are in the directory. If there's exactly one of each, we indicate success by returning the folder name. Anything else will result in a failure, which will be indicated by a string starting with `Error`:

```python
def ProcessProposalFolder(self, folder_path: str) -> str:
    xlsx_count = 0
    docx_count = 0

    for file in os.listdir(folder_path):
        if file.endswith(".xlsx"):
            xlsx_count += 1
        elif file.endswith(".docx"):
            docx_count += 1

    if xlsx_count == 1 and docx_count == 1:
        return "Success"
    elif xlsx_count == 0 and docx_count == 0:
        return "Error: No files found"
    elif xlsx_count == 0:
        return "Error: No Excel spreadsheet found"
    elif docx_count == 0:
        return "Error: No Word document found"
    else:
        return "Error: multiple files found"
```

Now that we have the function to kick off the pipeline, let's see what needs to be done with the plugins we wrote in *Chapter 3* to make them usable for this pipeline.

Modifying the Excel native plugin

We need to make a few changes to the Excel plugin we created for *Chapter 3*:

- Standardize the error message to always start with `Error`
- Standardize the success message to always return the folder
- At the beginning of every function, if the input starts with `Error`, do nothing and simply pass the received input forward

To save space, the following code shows the changes to only one of the functions, CheckTabs. The full modified code is available in the GitHub repository.

C#

```
    [KernelFunction, Description("Checks that the spreadsheet contains
the correct tabs, 2024 and 2025")]
    public string CheckTabs([Description("The file path to the
spreadsheet")] string folderPath)
    {
        if (folderPath.StartsWith("Error"))
        {
            return folderPath;
        }
```

The preceding code checks if the input received contains an error. Since this is going to be used in a pipeline, any errors in previous steps will be received here.

If we get an error, we're simply going to pass the error forward in the following code:

```
        string filePath = GetExcelFile(folderPath);
        try
        {
            FileInfo fileInfo = new FileInfo(filePath);

            if (!fileInfo.Exists)
            {
                return "Error: File does not exist.";
            }
```

Note that we are ensuring that any error message starts with Error, to make sure they're easy to detect when they are received by other pipeline functions.

We now check how many sheets are in the file:

```
            using (var package = new ExcelPackage(fileInfo))
            {
                ExcelPackage.LicenseContext = OfficeOpenXml.
LicenseContext.NonCommercial;
                var workbook = package.Workbook;
                if (workbook.Worksheets.Count != 2)
                {
                    return "Error: Spreadsheet does not contain 2
tabs.";
                }
                if (workbook.Worksheets.Any(sheet => sheet.Name ==
```

```
"2024") && workbook.Worksheets.Any(sheet => sheet.Name == "2025"))
                {
                        return folderPath;
                }
```

In the case of success, we simply pass the folder forward.

Otherwise, the following code passes the error forward:

```
                else
                {
                        return "Error: Spreadsheet does not contain 2024
and 2025 tabs.";
                }
            }
        }
        catch (Exception ex)
        {
            return $"Error: An error occurred: {ex.Message}";
        }
    }
```

Let's see the changes in Python. To save space, we're only showing the changes to the CheckTabs function. The full code for all functions is in the GitHub repository.

Python

```
        @kernel_function(
        description="Checks that the spreadsheet contains the correct
tabs, 2024 and 2025",
        name="CheckTabs",
    )
    def CheckTabs(self,
                    input: Annotated[str, "The path to the
spreadsheet"]) -> Annotated[str, "The result of the check"]:
        if path.startswith("Error"):
            return path
```

If the previous function sent us an error, we simply pass it forward.

Otherwise, we continue:

```
        try:
            filePath = self.GetExcelFile(path)
            workbook = openpyxl.load_workbook(filePath)
            sheet_names = workbook.sheetnames
```

```
        if sheet_names == ['2024', '2025']:
            return path
```

To indicate success, we simply return the folder we received as input:

```
        else:
            return "Error: the spreadsheet does not contain the
correct tabs"
    except Exception as e:
        return f"Error: an exception {e} occurred when trying to
open the spreadsheet"
```

Now that we have standardized the input and output of the functions, it makes it easier to extend the pipeline, since all the functions expect the same input, a folder, and when the function succeeds, it simply passes the folder to the next function.

In the case of an error, the string passed as an input will start with `Error`, and all functions will simply pass it forward. These changes make it easier to change the order of steps and to add or remove steps.

Now that we're done with the changes in the Excel plugin, let's make changes to the `ParseWordDocument` plugin that extracts text from Word documents.

Modifying the Word native plugin

The modifications to the Word native plugin are very simple. In *Chapter 3*, when we were calling the native plugin directly, we created a function called `ExtractTextUnderHeading` that received two parameters: a path to a file and the heading that we wanted to extract, and we called that function three times so that we could extract the text under the three headings we wanted.

In pipelines, since we can only have one parameter, we will create three functions named `ExtractTeam`, `ExtractExperience`, and `ExtractImplementation` that receive the folder as a parameter and call the `ExtractTextUnderHeading` function that we created in *Chapter 3* with the appropriate heading parameter, respectively `"Team"`, `"Experience"`, and `"Implementation"`.

Like we did with the Excel plugin, we will also make the following changes:

- Standardize the error message to always start with `Error`
- Standardize the success message to always return the folder
- At the beginning of every function, if the input starts with `Error`, do nothing and simply pass the received input forward

To save space, we only show one of the functions here. The full code, including the modifications in the `ExtractTextUnderHeading` function, is in the GitHub repository:

C#

```
    [KernelFunction, Description("Extracts the text under the Team
heading in the Word document")]
    public static string ExtractTeam(string folderPath)
    {
        if (folderPath.Contains("Error"))
        {
            return folderPath;
        }
        string text = ExtractTextUnderHeading(folderPath, "Team");
        return $"FolderPath: {folderPath}\n"  + text;


    }
```

Python

```
    def ExtractTeam(self, folder_path: str) -> str:
        if folder_path.startswith("Error"):
            return folder_path
        doc_path = self.get_first_docx_file(folder_path)
        text = self.ExtractTextUnderHeading(doc_path, 'Team')
        return f"FolderPath: {folder_path}\n{text}"
```

In both cases, in addition to creating three function wrappers for `ExtractTextUnderHeading` to extract the text, we also perform two more tasks. The first is to check whether the previous step of the pipeline sent an error message. If it did, we simply pass it on. The second is to prepend the folder path to the text in a line called `FolderPath`. This will be used in the semantic functions. When the semantic function decides that the text it read fulfills the requirements, it will return the folder path, as is expected by functions in the pipeline.

Let's work on the semantic functions.

Modifying the semantic functions

The main modification that we need to make to the semantic functions is to ensure they understand the inputs and provide the appropriate outputs – either the folder in the case of success or an error message in the case of failure.

One way to do this is to encode the input into tags and then tell the AI service to perform operations on the contents of the tag. There was no need to modify the config.json files, only the skprompt.txt files.

CheckTeamV2: skprompt.txt

```
=====
 {{$input}}
=====


Check the contents of the text between the ===== and =====.

If the contents are an error message, respond with the error message,
including the word "Error:" at the beginning.

Otherwise, the first line of the text between the ===== and ===== will
contain the FolderPath.
The other lines will contain the team's experience.

We require the team's experience to have at least one person with a
Ph.D. and at least one person with a degree in the medical sciences.

Think carefully about the team's experience.

If the team fulfills the requirements, your answer should be the
contents of the FolderPath field, nothing more.

If the team does not fulfill the requirements, reply with "Error: Team
does not meet requirements."
```

In this semantic function, we tell the AI to check the contents of the text between ===== for an error and simply pass it on if it finds one. If the contents between the ===== tags are not an error, they will contain the folder we're processing in a line starting with FolderPath and the text from the Team heading in the Word document. We tell the AI to return the folder if the team fulfills the requirements we list or to return an error message if they don't:

CheckDatesV2: skprompt.txt

```
=====
 {{$input}}
=====


Check the contents of the text between the tags ===== and =====.
```

```
If the contents are an error message, respond with the error message,
including the word "Error:" at the beginning.

Otherwise, the text between the tags ===== and ===== will contain
a line starting with FolderPath and the rest of the field will contain
a description of a timeline for vaccine implementation.

Think carefully about the following question:
Ignoring the FolderPath line, does the timeline contain dates?

If there are no dates listed, say "Error: No dates listed" .

If the dates are outside of the 2024-2025 range, say "Error: dates out
of range".

If there are dates between 2024 and 2025, respond only with the
contents of the FolderPath field, nothing more.
```

We now ask the AI to check the contents between the ===== tags for an error. The same as before, we simply pass it on if we find it. We then check whether the dates proposed for the vaccination campaign are within our expectations. If they are, we return the folder contained in the `FolderPath` line inside the tag. Otherwise, we return an error message:

CheckPreviousProjectV2: skprompt.txt

```
=====
 {{$input}}
=====

Check the contents of the text between the ===== and =====.

If the contents are an error message, respond with the error message,
including the word "Error:" at the beginning.

Otherwise, the text between the ===== and ===== will contain a line
starting with FolderPath and the rest of the field will contain a
description of the teams experience.

Ignoring the FolderPath line, does the description of the teams
experience
indicate they have enough experience to conduct a massive vaccination
campaign in a new country?

If they have had a successful experience in Atlantis or another large
country, respond only with the
```

```
contents of the FolderPath field, nothing more.

Otherwise, respond with "Error: Not enough experience."
```

The final semantic function is very similar to the previous two. We ask the AI to check the contents between the ===== tags and return an error message if the team requesting for funding does not have enough experience as demonstrated by a previous project.

Now that we're done with all the steps of our process, let's assemble them into a pipeline and run it.

Creating and calling the pipeline

Calling the pipeline requires creating a kernel, loading it with the functions we want, and then calling them in sequence. Since we're going to use semantic functions, we also need to add an AI service to the kernel. Evaluating text against requirements can be a complex task, and therefore we will use GPT-4 to execute it. GPT 3.5 can work if the documents are simple, but some of our documents have more than one page, and that can be too much for GPT 3.5 to handle well.

C#

In the following code block, we load all the native and semantic plugins into our kernel:

```csharp
using Microsoft.SemanticKernel;
using Plugins.ProposalChecker;
using System;
using System.IO;

var (apiKey, orgId) = Settings.LoadFromFile();

var builder = Kernel.CreateBuilder();
builder.AddOpenAIChatCompletion("gpt-4", apiKey, orgId);
builder.Plugins.AddFromPromptDirectory("../../../plugins/
ProposalCheckerV2");
builder.Plugins.AddFromType<Helpers>();
builder.Plugins.AddFromType<ParseWordDocument>();
builder.Plugins.AddFromType<CheckSpreadsheet>();
var kernel = builder.Build();
```

Then, we create variables for each of the functions:

```csharp
KernelFunction processFolder = kernel.Plugins["Helpers"]
["ProcessProposalFolder"];
KernelFunction checkTabs = kernel.Plugins["CheckSpreadsheet"]
["CheckTabs"];
KernelFunction checkCells = kernel.Plugins["CheckSpreadsheet"]
```

```
["CheckCells"];
KernelFunction checkValues = kernel.Plugins["CheckSpreadsheet"]
["CheckValues"];
KernelFunction extractTeam = kernel.Plugins["ParseWordDocument"]
["ExtractTeam"];
KernelFunction checkTeam = kernel.Plugins["ProposalCheckerV2"]
["CheckTeamV2"];
KernelFunction extractExperience = kernel.Plugins["ParseWordDocument"]
["ExtractExperience"];
KernelFunction checkExperience = kernel.Plugins["ProposalCheckerV2"]
["CheckPreviousProjectV2"];
KernelFunction extractImplementation = kernel.
Plugins["ParseWordDocument"]["ExtractImplementation"];
KernelFunction checkDates = kernel.Plugins["ProposalCheckerV2"]
["CheckDatesV2"];
```

Creating variables for each of the functions is not strictly necessary – you could simply put the code on the right-hand side of each of the preceding assignments directly into the pipeline call.

For example, instead of

```
KernelFunctionCombinators.Pipe(new[] {
    processFolder,
    checkTabs}
```

You could write:

```
KernelFunctionCombinators.Pipe(new[] {
kernel.Plugins["ProposalCheckerV2"]["ProcessProposalFolder"]}
    kernel.Plugins["ProposalCheckerV2"]["CheckTabs"]}
```

Assigning it to variables makes the contents of the pipeline call much shorter, and that can make it easier to maintain.

Next, we create a pipeline with the `Pipe` method of `KernelFunctionCombinators`, simply listing the steps in the order we want them to be called:

```
KernelFunction pipeline = KernelFunctionCombinators.Pipe(new[] {
    processFolder,
    checkTabs,
    checkCells,
    checkValues,
    extractTeam,
    checkTeam, .
    extractExperience,
    checkExperience,
    extractImplementation,
```

```
        checkDates
}, "pipeline");
```

The next step will be to call the pipeline:

```
var proposals = Directory.GetDirectories("../../../data/proposals");

// print each directory
foreach (var proposal in proposals)
{
    // convert to absolute path
    string absolutePath = Path.GetFullPath(proposal);

    Console.WriteLine($"Processing {absolutePath}");
    KernelArguments context = new() { { "folderPath", absolutePath }
};
    string result = await pipeline.InvokeAsync<string>(kernel,
context);
    Console.WriteLine(result);
    if (result == absolutePath)
    {
        Console.WriteLine("Success!");
    }

    Console.WriteLine();
}
```

We get the path of the `data/proposals` folder that contains our proposals. Each proposal is a subfolder of that folder. We iterate over each of the subfolders of the `data/proposal` folder and call our pipeline. If we don't find any errors, we print `Success`. Otherwise, we list the errors we found.

Python

We start by creating our kernel, adding the GPT-4 service to it, and adding all the native and semantic plugins to it:

```
import asyncio
from semantic_kernel.connectors.ai.open_ai import OpenAIChatCompletion
import semantic_kernel as sk
from CheckSpreadsheet import CheckSpreadsheet
from ParseWordDocument import ParseWordDocument
from Helpers import Helpers
import os
async def pipeline(kernel, function_list, input):
    for function in function_list:
```

```
        args = KernelArguments(input=input)
        input = await kernel.invoke(function, args)
    return input

async def main():
    kernel = sk.Kernel()
    api_key, org_id = sk.openai_settings_from_dot_env()
    gpt4 = OpenAIChatCompletion("gpt-4", api_key, org_id)
    kernel.add_chat_service("gpt4", gpt4)

    parse_word_document = kernel.import_skill(ParseWordDocument())
    check_spreadsheet = kernel.import_skill(CheckSpreadsheet())
    helpers = kernel.import_skill(Helpers())
    interpret_document = kernel.import_semantic_skill_from_
directory("../../plugins", "ProposalCheckerV2")
```

Note that we added the pipeline function that we created in the previous section.

The final step is to create a function list and call the pipeline for each document:

```
    data_path = "../../data/proposals/"

    for folder in os.listdir(data_path):
        if not os.path.isdir(os.path.join(data_path, folder)):
            continue
        print(f"\n\nProcessing folder: {folder}")
        function_list = [
            helpers['ProcessProposalFolder'],
            check_spreadsheet['CheckTabs'],
            check_spreadsheet['CheckCells'],
            check_spreadsheet['CheckValues'],
            parse_word_document['ExtractTeam'],
            interpret_document['CheckTeamV2'],
            parse_word_document['ExtractExperience'],
            interpret_document['CheckPreviousProjectV2'],
            parse_word_document['ExtractImplementation'],
            interpret_document['CheckDatesV2']
        ]
        process_result = await pipeline(kernel, function_list,
os.path.join(data_path, folder))

        result = (str(process_result))
        if result.startswith("Error"):
            print(result)
            continue
```

```
    else:
        print("Success")
```

The full pipeline obtains the expected results, with the proposals that fulfill all the requirements returning success and the proposals with problems returning an error message describing the problem.

The results are displayed here:

```
Processing folder: correct
Success

Processing folder: incorrect01
Error: the spreadsheet does not contain the correct tabs

Processing folder: incorrect02
Error: Sum of values in year 2025 exceeds 1,000,000.

Processing folder: incorrect03
Error: More than 10% growth found from B2 to B3 in sheet 2024.

Processing folder: incorrect04
Error: non-numeric inputs

Processing folder: incorrect05
Error: No Word document found

Processing folder: incorrect06
Error: No Excel spreadsheet found

Processing folder: incorrect07
Error: Not enough experience.

Processing folder: incorrect08
Error: Team does not meet requirements.

Processing folder: incorrect09
Error: dates out of range

Processing folder: incorrect10
Error: multiple files found
```

Summary

Before having the help of AI, reading and interpreting a document required using the time of a person, or writing a specialized machine learning model. Semantic Kernel allows you to write code to analyze large and complex documents.

In our pipeline, the `CheckSpreadsheet` native plugin does not strictly require Semantic Kernel and could be done in a separate step, since it only runs code that is never read by AI. We added it to the pipeline to make our end-to-end solution more streamlined.

The `ParseWordDocument` native plugin, on the other hand, helps Semantic Kernel receive the information in parts. Breaking the document into parts makes the semantic functions simpler: each function can evaluate just a portion of the document. For example, the function that evaluates the *Teams* section of the document just needs to check the team qualifications. That makes the function a lot simpler to write than a function that reads the whole document and decides about all sections of the document in a single step.

The real value that AI adds to this process, therefore, is in the semantic plugin. The tasks of evaluating sections of the document that are implemented by the semantic functions in the `ProposalCheckerV2` plugin are the ones that would formerly require either a lot of human effort or a specialized machine learning model. This chapter showed how to execute these tasks just by describing what the requirements were in three short `skprompt.txt` files.

In this chapter, we created our pipeline manually, explicitly naming the functions that we wanted to call and in which order. In the next chapter, we will learn how to use a planner. The planner will receive the user request and decide which functions to call and in which order.

References

[1] A. S. Luccioni, Y. Jernite, and E. Strubell, "Power Hungry Processing: Watts Driving the Cost of AI Deployment?" arXiv, Nov. 28, 2023. doi: 10.48550/arXiv.2311.16863.

5

Programming with Planners

In the previous chapter, we learned how to chain functions manually to perform complex tasks. In this chapter, we will learn how to use **planners** to chain functions automatically. Chaining functions automatically can give your users a lot of flexibility, allowing them to use your application in ways that you don't have to write code for.

In this chapter, we will learn how planners work, when to use them, and what to be careful about. We will also learn how to write functions and build a kernel that helps planners build good plans.

In this chapter, we'll be covering the following topics:

- What a planner is and when to use one
- Creating and using a planner to run a simple function
- Designing functions to help a planner decide the best way to combine them
- Using a planner to allow users combine functions in complex ways without having to write code

By the end of the chapter, you will have learned how to empower users by giving them the ability to make requests in natural language, allowing them to solve complex problems that you didn't have to write code for.

Technical requirements

To complete this chapter, you will need to have a recent, supported version of your preferred Python or C# development environment:

- For Python, the minimum supported version is Python 3.10, and the recommended version is Python 3.11
- For C#, the minimum supported version is .NET 8

In this chapter, we will call OpenAI services. Given the amount that companies spend on training these LLMs, it's no surprise that using these services is not free. You will need an **OpenAI API** key, either directly through **OpenAI** or **Microsoft**, via the **Azure OpenAI** service.

If you are using .NET, the code for this chapter is at `https://github.com/PacktPublishing/` `Building-AI-Applications-with-Microsoft-Semantic-Kernel/tree/main/` `dotnet/ch5`.

If you are using Python, the code for this chapter is at `https://github.com/PacktPublishing/` `Building-AI-Applications-with-Microsoft-Semantic-Kernel/tree/main/` `python/ch5`.

You can install the required packages by going to the GitHub repository and using the following: `pip install -r requirements.txt`.

What is a planner?

So far, we have been performing complex requests by performing the function calls ourselves. This, however, requires you to restrict the kind of requests that your users can make to what you can predict and write ahead of time. It also restricts your users to only generating one output at a time. Sometimes, you may want to give them the ability to do more.

For example, if you have a semantic function that allows users to request jokes (as we built in *Chapter 1*) and a user requests "*tell me a knock-knock joke*," you can simply call the semantic function that tells knock-knock jokes. But if the user requests three knock-knock jokes, the function wouldn't know how to handle it.

A planner is a built-in function from Semantic Kernel that receives a user request and then goes through the descriptions of all the functions, parameters, and outputs of the functions you loaded in your kernel and decides the best way to combine them, generating a **plan**.

At the time of writing, there are two planners – a **Handlebars planner** and a **Function Calling Stepwise planner**, which we will call a Stepwise planner for short. Both are used in the same way, but internally, they work in different ways. When Semantic Kernel uses the Handlebars planner, it asks the AI service (for example, GPT-3.5 or GPT-4) to write code that will call the functions you loaded into the kernel in a scripting language called Handlebars. The Handlebars planner is very new and still experimental. It is expected to consume fewer tokens than the Stepwise planner, as programming languages can be more efficient in expressing complex ideas such as conditionals and loops. The Stepwise planner generates a plan that is a dialog with a chat service, which can be longer than the plan generated by the Handlebars planner and consume more tokens. Currently, one of the major limitations of the Handlebars planner is that it is only available in C#, although a Python version is likely to be released in 2024.

To understand better how a planner works, assume you have a plugin that generates stories, a plugin that breaks stories into small parts, and a plugin that generates images. You load all these plugins into the kernel. The user submits a request:

"Create a two-page story about a data scientist that solves crimes with his faithful canine companion, break it into small parts, and generate an image in the style of Frank Miller for each part."

The planner will go through the functions you loaded in the kernel and determine the best order to call them, automatically producing a storyboard without you having to write any additional code other than the initial plugins.

Planners can enable your users to execute complex tasks with minimal effort from your side. Let's see when to use them.

When to use a planner

Planners can help you as a developer in two ways:

- Users can combine functions of your application in ways that you didn't think of. If you make the functions of your application available as atomic functions inside plugins and give the users the ability to make requests to a planner, then the planner can combine these atomic functions in workflows without you having to write any code.

- As AI models improve, planners get better without you having to write any additional code. When Semantic Kernel was initially designed, the best AI model available was GPT-3.5 Turbo. Since then, we have had the releases of GPT-4 and GPT-4 Turbo, both with more capabilities. An application built with Semantic Kernel that used GPT-3.5 Turbo can now use GPT-4 Turbo with a minor configuration change.

There are, however, some considerations when using planners:

- **Performance**: Planners need to read all the functions in your kernel and combine them with the user request. The richer your kernel is, the more functionality you can give to your users, but it will take longer for the planner to go through all the descriptions and combine them. In addition, newer models such as GPT-4 generate better plans, but they are slower, and future models could be even slower. You need to find a good balance between the number of functions you make available to your users and the models you use. When testing your application, if you find that the planner delay is noticeable, you will also need to incorporate UI cues into your application so that users know that something is happening.

- **Cost**: Generating a plan can consume many tokens. If you have lots of functions, and the user request is complex, Semantic Kernel will need to submit a very long prompt to the AI service containing the descriptions of the functions available in your kernel, their inputs, and their outputs to the AI service, in addition to the user request. The generated plan may also be long, and the AI service will bill you for the cost of both the submitted prompt and the output. One way to avoid this is monitoring which requests users create frequently and saving plans for those so that they don't have to be regenerated every time. Note, however, that if you save plans and there's an upgrade in the backend model (for example, GPT-5 is launched), you have to remember to re-generate these plans to take advantage of the new model's capabilities.

- **Testing**: Using a planner makes testing your application a lot harder. For example, it's possible that your kernel has so many functions and that the user requests can be so complex that the planner will exceed the context window of the model you are using. You will need to do something to handle this runtime error, such as limiting the size of the user requests or the number of functions available in your kernel. In addition, while the planner works most of the time, it's possible that the planner will occasionally produce faulty plans, such as plans that hallucinate functions. You will need to provide error handling for that. Interestingly, in practice, the simple error-handling technique of just resubmitting the failed plan, telling the AI service that the plan didn't work, and asking "*can you fix it?*" usually works.

With all that in mind, let's see how to use the planner. The first step is to instantiate a planner.

Instantiating a planner

Instantiating and using a planner is straightforward. In C#, we are going to use the Handlebars planner, and in Python, we will use the Stepwise planner.

C#

C# includes the new `HandlebarsPlanner`, which allows you to create plans that include loops, making them shorter. Before using the Handlebars planner in C#, you need to install it with the following command:

```
dotnet add package Microsoft.SemanticKernel.Planners.Handlebars
--prerelease
```

To configure your Handlebars planner, you will also need to install the OpenAI planner connector with the following:

```
dotnet add package Microsoft.SemanticKernel.Planners.OpenAI
--prerelease
```

Note that the planner is experimental, and C# will give you an error unless you let it know that you are OK with using experimental code, by adding a `pragma` directive to your code:

```
#pragma warning disable SKEXP0060
```

To create a planner, we execute the following code:

```
var plannerOptions = new HandlebarsPlannerOptions()
    {
        ExecutionSettings = new OpenAIPromptExecutionSettings()
        {
            Temperature = 0.0,
            TopP = 0.1,
            MaxTokens = 4000
        },
        AllowLoops = true
    };
var planner = new HandlebarsPlanner(plannerOptions);
```

Microsoft recommends using a low `Temperature` and `TopP` for your planners, minimizing the chance of the planner creating non-existent functions. Planners may consume lots of tokens; therefore, we usually set `MaxTokens` to a high value to avoid having a runtime error.

Now, let's see how to create a planner in Python.

Python

In Python, the Handlebar planner is not available yet, so we need to instantiate the Stepwise planner. Plans created by the Stepwise planner tend to be longer than Handlebars plans. To add the Stepwise planner to your Python project, you need to import the `FunctionCallingStepwisePlanner` and `FunctionCallingStepwisePlannerOptions` classes from the `semantic_kernel.planners` package:

```
from semantic_kernel.planners import FunctionCallingStepwisePlanner,
FunctionCallingStepwisePlannerOptions
import semantic_kernel as sk
```

It's usually a good idea to give planners plenty of tokens. The following is a sample command to create a planner, assuming you loaded a service with `service_id` set to `gpt4` in your semantic kernel:

```
planner_options = FunctionCallingStepwisePlannerOptions(
        max_tokens=4000,
    )
planner = FunctionCallingStepwisePlanner(service_id="gpt4",
options=planner_options)
```

Now, let's create and run a plan for a user request.

Creating and running a plan

Now that we have a planner, we can use it to create a plan for a user's request and then invoke the plan to get a result. In both languages, we use two steps, one to create the plan and another one to execute it.

For the next two code snippets, assume you have the user's request loaded into the `ask` string. Let's see how to call the planner:

C#

```
var plan = await planner.CreatePlanAsync(kernel, ask);
var result = await plan.InvokeAsync(kernel);
Console.Write ($"Results: {result}");
```

Python

```
result = await planner.invoke(kernel, ask)
print(result.final_answer)
```

You may remember from *Chapter 1* that in Python, the result variable contains all the steps to create the plan, so in order to see the plan's results, you need to print `result.final_answer`. If you print the `result` variable, you'll get a large JSON object.

An example of how a planner can help

Let's see a simple example that already shows how a planner can help. Let's say you create an application that helps aspiring comedians create jokes. You create and connect it to the `jokes` semantic plugin that we created in *Chapter 1*. That plugin contains a semantic function that creates knock-knock jokes.

You can create a UI that allows users to enter a theme (say, "*dog*") and call that function to create a knock-knock joke. If the user wants to create 100 jokes, they'll need to use that UI 100 times. You can work around that problem by creating yet another UI that asks for the number of jokes the user wants to create. However, if the user wants to create multiple jokes for multiple themes, then they must use your two UIs for each theme they want to create a joke for.

Conversely, with just the semantic function and the planner, you can allow your user to describe what they want in natural language, such as the following:

"*Create four knock-knock jokes – two about dogs, one about cats, and one about ducks.*"

The complete code is as follows:

C#

```
#pragma warning disable SKEXP0060

using Microsoft.SemanticKernel;
using Microsoft.SemanticKernel.Planning.Handlebars;
using Microsoft.SemanticKernel.Connectors.OpenAI;

var (apiKey, orgId) = Settings.LoadFromFile();

var builder = Kernel.CreateBuilder();
builder.AddOpenAIChatCompletion("gpt-4", apiKey, orgId);
var kernel = builder.Build();

var pluginsDirectory = Path.Combine(System.IO.Directory.
GetCurrentDirectory(),
        "..", "..", "..", "plugins", "jokes");

kernel.ImportPluginFromPromptDirectory(pluginsDirectory);

var plannerOptions = new HandlebarsPlannerOptions()
    {
        ExecutionSettings = new OpenAIPromptExecutionSettings()
        {
            Temperature = 0.0,
            TopP = 0.1,
            MaxTokens = 4000
        },
        AllowLoops = true
    };

var planner = new HandlebarsPlanner(plannerOptions);
var ask = "Tell four knock-knock jokes: two about dogs, one about cats
and one about ducks";
var plan = await planner.CreatePlanAsync(kernel, ask);
var result = await plan.InvokeAsync(kernel);
Console.Write ($"Results: {result}");
```

Python

```python
import asyncio
from semantic_kernel.connectors.ai.open_ai import OpenAIChatCompletion
from semantic_kernel.planners import FunctionCallingStepwisePlanner,
FunctionCallingStepwisePlannerOptions
from semantic_kernel.utils.settings import openai_settings_from_dot_
env
import semantic_kernel as sk
from dotenv import load_dotenv

async def main():
    kernel = sk.Kernel()
    api_key, org_id = openai_settings_from_dot_env()

    gpt35 = OpenAIChatCompletion("gpt-3.5-turbo", api_key, org_id,
service_id = "gpt35")
    gpt4 = OpenAIChatCompletion("gpt-4", api_key, org_id, service_id =
"gpt4")

    kernel.add_service(gpt35)
    kernel.add_service(gpt4)
    kernel.add_plugin(None, plugin_name="jokes", parent_
directory="../../plugins/")

    planner_options = FunctionCallingStepwisePlannerOptions(
        max_tokens=4000,
    )
    planner = FunctionCallingStepwisePlanner(service_id="gpt4",
options=planner_options)

    prompt = "Create four knock-knock jokes: two about dogs, one about
cats and one about ducks"
    result = await planner.invoke(kernel, prompt)

    print(result.final_answer)

if __name__ == "__main__":
    asyncio.run(main())
```

In the preceding code, we created our kernel and added our jokes plugin to it. Now, let's create the planner.

Results

You will get the following results for both Python and C#:

```
1st Joke: Knock, knock!
Who's there?
Dog!
Dog who?
Dog who hasn't barked yet because he doesn't want to interrupt this
hilarious joke!
2nd Joke: Knock, knock!
Who's there?
Dog!
Dog who?

Dog who forgot his keys, let me in!
3rd Joke: Knock, knock!
Who's there?
cat!
cat who?
Cat-ch me if you can, I'm the gingerbread man!
4th Joke: Knock, knock!
Who's there?
Duck!
Duck who?

Duck down, I'm throwing a pie!
```

Note that with a single user request and a single call to `invoke`, Semantic Kernel generated several responses, without you having to write any loop, create any additional UIs, or chain any functions yourself.

Let's see what happens behind the scenes.

How do planners work?

Behind the scenes, the planner uses an LLM prompt to generate a plan. As an example, you can see the prompt that is used by `HandlebarsPlanner` by navigating to its prompt file in the Semantic Kernel repository, located at `https://github.com/microsoft/semantic-kernel/blob/7c3a01c1b6a810677d871a36a9211cca0ed7fc4d/dotnet/src/Planners/Planners.Handlebars/Handlebars/CreatePlanPrompt.handlebars`.

The last few lines of the prompt are the most important to understand how the planner works:

```
## Start
Now take a deep breath and accomplish the task:
1. Keep the template short and sweet. Be as efficient as possible.
2. Do not make up helpers or functions that were not provided to
you, and be especially careful to NOT assume or use any helpers or
operations that were not explicitly defined already.
3. If you can't fully accomplish the goal with the available helpers,
just print "{{insufficientFunctionsErrorMessage}}".
4. Always start by identifying any important values in the goal.
Then, use the `\{{set}}` helper to create variables for each of these
values.
5. The template should use the \{{json}} helper at least once to
output the result of the final step.
6. Don't forget to use the tips and tricks otherwise the template will
not work.
7. Don't close the ``` handlebars block until you're done with all the
steps.
```

The preceding steps define the set of rules that the planner uses to generate a plan using Handlebars.

Also, inside the prompt is what we call the **function manual** – that is, the instructions that the LLM will use to convert functions loaded into the kernel into text descriptions that are suitable for an LLM prompt:

```
{{#each functions}}
### `{{doubleOpen}}{{PluginName}}{{../nameDelimiter}}{{Name}}
{{doubleClose}}`
Description: {{Description}}
Inputs:
  {{#each Parameters}}
    - {{Name}}:
    {{~#if ParameterType}} {{ParameterType.Name}} -
    {{~else}}
        {{~#if Schema}} {{getSchemaTypeName this}} -{{/if}}
    {{~/if}}
    {{~#if Description}} {{Description}}{{/if}}
    {{~#if IsRequired}} (required){{else}} (optional){{/if}}
  {{/each}}
Output:
{{~#if ReturnParameter}}
  {{~#if ReturnParameter.ParameterType}} {{ReturnParameter.
ParameterType.Name}}
  {{~else}}
    {{~#if ReturnParameter.Schema}} {{getSchemaReturnTypeName
ReturnParameter}}
```

```
      {{else}} string{{/if}}
   {{~/if}}
   {{~#if ReturnParameter.Description}} - {{ReturnParameter.
Description}}{{/if}}
{{/if}}
{{/each}}
```

In summary, the planner is just a plugin that uses an AI service to translate a user request into a series of callable function steps, and then it generates the code that calls these functions, returning the result.

To decide which functions to call and how to call them, planners rely on the descriptions you wrote for the plugin. For native functions, the descriptions are in function decorators, while for semantic functions, they are in the `config.json` file.

Planners will send your descriptions to an AI service as part of a prompt, with instructions that tell the AI service how to combine your descriptions into a plan. Writing good descriptions can help the AI service to create better plans.

Here are some things you should do:

- **State whether inputs are required**: If a function requires an input, you should state that in the input's description by using `required=true` so that the model knows to provide an input. If you don't do that, the created plan may not include a required parameter and will fail when executing.

- **Provide examples**: Your description can provide examples of how to use the function and what the acceptable inputs and outputs are. For example, if you have a function that turns lights on in a location with the description "*Location where the lights should be turned on*," and the location must be the kitchen or the garage, you can add "*The location must be either 'kitchen' or 'garage'*" to the description. With that extra description, the planner will know not to call that function if the user asks to "*turn everything on in the bedroom.*"

Here are some things to avoid:

- **Short descriptions**: If your function, inputs, or output descriptions are very short, it's possible that they are not going to convey enough information to the planner about the context where they would be used. For example, it's better to say that the output of a function is "*a knock-knock joke that follows a theme*" than "*joke.*"

- **Very long descriptions**: Remember that the descriptions will be submitted as part of a prompt that will incur costs. If your description is very long (for example, you provide three examples for every function), you will pay for it. Make sure that what you write in the descriptions is close to what's necessary.

- **Conflicting descriptions**: If many of your functions have similar or the same description, the planner can get confused. For example, imagine that you create a jokes plugin that can create different types of jokes (knock-knock jokes, puns, absurdist jokes, etc.) but the description of all the functions is simply "*creates a joke*." The planner will not know which function to call because the description tells it that all functions do the same thing.

If you are not getting the results that you expect when you use the planner, the first place you should look is in the descriptions you wrote for the functions, their inputs, and their outputs. Usually, just improving the descriptions a little helps the planner a lot. Another solution is to use a newer model. For example, if the plans are failing when you use GPT-3.5 and you already checked the descriptions, you may consider testing GPT-4 and seeing whether the results improve substantially.

Let's see a comprehensive example.

Controlling home automation with the planner

To get a better idea of what the planner can do, we will create a home automation application. We will not actually write functions that really control home automation, but assuming those exist, we will write their wrappers as native functions. We will also add a semantic function to our kernel and incorporate it into the planner.

We assume that we have a house with four rooms – a garage, kitchen, living room, and bedroom. We have automations to operate our garage door, operate the lights in all rooms, open the windows in the living room and in the bedroom, and operate the TV.

Since our objective is to learn about Semantic Kernel and not about home automation, these functions will be very simple. We want our user to be able to say something such as "*turn on the lights of the bedroom*," and the result will be that our native function will say "*bedroom lights turned on.*"

The power of using the planner is shown when a user makes requests that require multiple steps, such as "*turn off the bedroom light and open the window*," or even something more complex, such as "*turn off the living room lights and put on a highly rated horror movie on the TV.*"

Creating the native functions

We will start by creating four native functions for home automation, one to operate the lights, one to operate the windows, one to operate the TV, and one to operate the garage door:

C#

```
using System.ComponentModel;
using Microsoft.SemanticKernel;
public class HomeAutomation
{
```

```
    [KernelFunction, Description("Turns the lights of the living room,
kitchen, bedroom or garage on or off.")]
    public string OperateLight(
        [Description("Whether to turn the lights on or off. Must be
either 'on' or 'off'")] string action,
        [Description("The location where the lights must be turned on
or off. Must be 'living room', 'bedroom', 'kitchen' or 'garage'")]
string location)
    {
        string[] validLocations = {"kitchen", "living room",
"bedroom", "garage" };
        if (validLocations.Contains(location))
        {
            string exAction = $"Changed status of the {location}
lights to {action}.";
            Console.WriteLine(exAction);
            return exAction;
        }
        else
        {
            string error = $"Invalid location {location} specified.";
            return error;
        }
    }
```

The most important parts of the function are the `Description` decorators for the function itself and the parameters. They are the ones that the planner will read to learn how to use the function. Note that the descriptions specify what the valid parameters are. This helps the planner decide what to do when it receives an instruction for all locations.

The function just verifies that the location is valid and prints the action that the home automation would have taken if it were real.

The other functions simply repeat the same preceding template for their objects (the window, TV, and garage door);

```
    [KernelFunction, Description("Opens or closes the windows of the
living room or bedroom.")]
    public string OperateWindow(
        [Description("Whether to open or close the windows. Must be
either 'open' or 'close'")] string action,
        [Description("The location where the windows are to be opened
or closed. Must be either 'living room' or 'bedroom'")] string
location)
    {
        string[] validLocations = {"living room", "bedroom"};
        if (validLocations.Contains(location))
```

```
        {
            string exAction = $"Changed status of the {location}
windows to {action}.";
            Console.WriteLine(exAction);
            return exAction;
        }
        else
        {
            string error = $"Invalid location {location} specified.";
            return error;
        }
    }

    [KernelFunction, Description("Puts a movie on the TV in the living
room or bedroom.")]
    public string OperateTV(
        [Description("The movie to play on the TV.")] string movie,
        [Description("The location where the movie should be played
on. Must be 'living room' or 'bedroom'")] string location)
    {
        string[] validLocations = {"living room", "bedroom"};
        if (validLocations.Contains(location))
        {
            string exAction = $"Playing {movie} on the TV in the
{location}.";
            Console.WriteLine(exAction);
            return exAction;
        }
        else
        {
            string error = $"Invalid location {location} specified.";
            return error;
        }
    }

    [KernelFunction, Description("Opens or closes the garage door.")]
    public string OperateGarageDoor(
        [Description("The action to perform on the garage door. Must
be either 'open' or 'close'")] string action)
    {
        string exAction = $"Changed status of the garage door to
{action}.";
        Console.WriteLine(exAction);
        return exAction;
```

```
            }
    }
```

Python

```python
from typing_extensions import Annotated
from semantic_kernel.functions.kernel_function_decorator import
kernel_function


class HomeAutomation:
    def __init__(self):
        pass

    @kernel_function(
        description="Opens or closes the windows of the living room or
bedroom.",
        name="OperateWindow",
    )
    def OperateWindow(self,
            location: Annotated[str, "The location where the
windows are to be opened or closed. Must be either 'living room' or
'bedroom'"],
            action: Annotated[str, "Whether to open or close the
windows. Must be either 'open' or 'close'"]) \
                -> Annotated[str, "The action that was performed on
the windows."]:
        if location in ["living room", "bedroom"]:
            action = f"Changed status of the {location} windows to
{action}."
            print(action)
            return action
        else:
            error = f"Invalid location {location} specified."
            return error
```

The preceding function is straightforward, checking that the location passed as a parameter is valid and printing what the automation would have done.

The most important parts of the function are the descriptions for inside the `kernel_function` and for each of the `Annotated` parameters, as the descriptions are what the planner will use to decide what to do.

Note that the descriptions specify what the valid parameters are. This helps the planner decide what to do when it receives a request to perform an action for all locations.

Now, let's create the other functions, following a similar structure:

```python
    @kernel_function(
        description="Turns the lights of the living room, kitchen,
bedroom or garage on or off.",
        name="OperateLight",
    )
    def OperateLight(self,
            location: Annotated[str, "The location where the lights
are to be turned on or off. Must be either 'living room', 'kitchen',
'bedroom' or 'garage'"],
            action: Annotated[str, "Whether to turn the lights on or
off. Must be either 'on' or 'off'"])\
                -> Annotated[str, "The action that was performed on
the lights."]:
        if location in ["kitchen", "living room", "bedroom",
"garage"]:
            action = f"Changed status of the {location} lights to
{action}."
            print(action)
            return action
        else:
            error = f"Invalid location {location} specified."
            return error


    @kernel_function(
        description="Puts a movie on the TV in the living room or
bedroom.",
        name="OperateTV",
    )
    def OperateTV(self,
            movie: Annotated[str, "The movie to play on the TV."],
            location: Annotated[str, "The location where the movie
should be played on. Must be 'living room' or 'bedroom'"]
            )\
                -> Annotated[str, "The action that was performed on
the TV."]:
        if location in ["living room", "bedroom"]:
            action = f"Playing {movie} on the TV in the {location}."
            print(action)
            return action
        else:
            error = f"Invalid location {location} specified."
            return error


    @kernel_function(
```

```
        description="Opens or closes the garage door.",
        name="OperateGarageDoor"
    )
    def OperateGarageDoor(self,
            action: Annotated[str, "The action to perform on the
garage door. Must be either 'open' or 'close'"])\
                -> Annotated[str, "The action that was performed on
the garage door."]:
        action = f"Changed the status of the garage door to {action}."
        print(action)
        return action
```

Now that we're done with native functions, let's add a semantic function.

Adding a semantic function to suggest movies

In addition to creating the preceding native functions that control different components of the house, we are also going to create a semantic function to suggest movies based on what the user requests. Semantic functions allow the user to make requests that require the use of an AI service – for example, to find the name of a movie based on a description or the name of an actor. You'll see that planners can seamlessly combine semantic and native functions.

As is always the case, the semantic function is the same for both C# and Python, but we need to carefully configure the skprompt.txt and config.json files to help the planner find the function and understand how to use it.

We start by creating a prompt:

skprompt.txt

The prompt is very simple, and it simply asks for a suggestion for a movie. To make things easier for the planner, the prompt specifies that GPT should only respond with the title of the movie, as well as what to do if the user already knows the movie they want to watch:

```
Given the request below, suggest exactly one movie that you think the
requestor is going to like. If the request is already a movie title,
just return that movie title.

Respond only with the title of the movie, nothing else.

Request:
{{ $input }}
```

Now, let's see the configuration file:

config.json

Here, it's again very important to fill in the `description` fields with as much detail as possible, as they are what the planner will use to decide which functions to call:

```json
{
    "schema": 1,
    "name": "RecommendMovie",
    "type": "completion",
    "execution_settings": {
        "default": {
            "temperature": 0.8,
            "number_of_responses": 1,
            "top_p": 1,
            "max_tokens": 4000,
            "presence_penalty": 0.0,
            "frequency_penalty": 0.0
        }
    },
    "input_variables": [
        {
            "name": "input",
            "description": "name or description of a movie that the
user wants to see. ",
            "required": true
        }
    ]
}
```

Now that all the native and semantic functions are configured, let's call the planner and see what it can do.

Invoking the planner

Once you load the kernel with all these functions, all you need to do is invoke the planner and pass the user request to it.

We are going to make four requests to the planner:

- *Turn on the lights in the kitchen*
- *Open the windows of the bedroom, turn the lights off, and put on The Shawshank Redemption on the TV*

- *Close the garage door and turn off the lights in all the rooms*

- *Turn off the lights in all rooms and play a movie in which Tom Cruise is a lawyer, in the living room*

Using the existing plugins, the planner will take care of everything that is needed to fulfill these requests. For example, to fulfill the last request, the planner needs to call the `OperateLight` native function for each of the four rooms and ask GPT for a recommendation of a movie in which Tom Cruise is a lawyer, which will likely be *A Few Good Men* or *The Firm*. The planner will automatically call the functions and simply provide the results.

Python

The core part of the code is to create and execute the plan, using `create_plan` and `invoke_async`, and then print the results:

```python
from semantic_kernel.connectors.ai.open_ai import OpenAIChatCompletion
from semantic_kernel.planning.stepwise_planner import StepwisePlanner
import semantic_kernel as sk
from HomeAutomation import HomeAutomation
from dotenv import load_dotenv
import asyncio

async def fulfill_request(planner: StepwisePlanner, request):
    print("Fulfilling request: " + request)

    variables = sk.ContextVariables()
    plan = planner.create_plan(request)
    result = await plan.invoke_async(variables)
    print(result)
    print("Request completed.\n\n")
```

Then, in the main function, we load the native functions and the semantic function in the kernel. This will make them available to the planner:

```python
async def main():
    kernel = sk.Kernel()
    api_key, org_id = sk.openai_settings_from_dot_env()
    gpt4 = OpenAIChatCompletion("gpt-4", api_key, org_id)
    kernel.add_chat_service("gpt4", gpt4)
    planner = StepwisePlanner(kernel)
    kernel.import_skill(HomeAutomation())
    kernel.import_semantic_skill_from_directory("../plugins/
MovieRecommender", "RecommendMovie")
```

After loading the function, we simply call `fulfill_request`, which will create and execute a plan for each `ask`:

```
    await fulfill_request(kernel, planner, "Turn on the lights in the
kitchen")
    await fulfill_request(kernel, planner, "Open the windows of the
bedroom, turn the lights off and put on Shawshank Redemption on the
TV.")
    await fulfill_request(kernel, planner, "Close the garage door and
turn off the lights in all rooms.")
    await fulfill_request(kernel, planner, "Turn off the lights in all
rooms and play a movie in which Tom Cruise is a lawyer in the living
room.")

if __name__ == "__main__":
    load_dotenv()
    asyncio.run(main())
```

C#

We start by creating a kernel and adding all the native functions and the semantic function we created for it. This will make these functions available to the planner:

```
using Microsoft.SemanticKernel;
using Microsoft.SemanticKernel.Planning.Handlebars;
#pragma warning disable SKEXP0060

var (apiKey, orgId) = Settings.LoadFromFile();
var builder = Kernel.CreateBuilder();
builder.AddOpenAIChatCompletion("gpt-4", apiKey, orgId);
builder.Plugins.AddFromType<HomeAutomation>();
builder.Plugins.AddFromPromptDirectory("../../../plugins/
MovieRecommender");
var kernel = builder.Build();
```

We then create a function that receives a `planner` and an `ask`, creating and executing a plan to fulfill that request:

```
void FulfillRequest(HandlebarsPlanner planner, string ask)
{
    Console.WriteLine($"Fulfilling request: {ask}");
    var plan = planner.CreatePlanAsync(kernel, ask).Result;
    var result = plan.InvokeAsync(kernel, []).Result;
    Console.WriteLine("Request complete.");
}
```

The last step is to create the planner and call the `FulfillRequest` function we created for each ask:

```
var plannerOptions = new HandlebarsPlannerOptions()
    {
        ExecutionSettings = new OpenAIPromptExecutionSettings()
        {
            Temperature = 0.0,
            TopP = 0.1,
            MaxTokens = 4000
        },
        AllowLoops = true
    };

var planner = new HandlebarsPlanner(plannerOptions);

FulfillRequest(planner, "Turn on the lights in the kitchen");
FulfillRequest(planner, "Open the windows of the bedroom, turn the
lights off and put on Shawshank Redemption on the TV.");
FulfillRequest(planner, "Close the garage door and turn off the lights
in all rooms.");
FulfillRequest(planner, "Turn off the lights in all rooms and play a
movie in which Tom Cruise is a lawyer in the living room.");
```

Note that the code that uses the planner was very short. Let's see the results:

```
Fulfilling request: Turn on the lights in the kitchen
Changed status of the kitchen lights to on.
Request complete.
Fulfilling request: Open the windows of the bedroom, turn the lights
off and put on Shawshank Redemption on the TV.
Changed status of the bedroom windows to open.
Changed status of the bedroom lights to off.
Playing Shawshank Redemption on the TV in the bedroom.
Request complete.
Fulfilling request: Close the garage door and turn off the lights in
all rooms.
Changed status of the garage door to close.
Changed status of the living room lights to off.
Changed status of the bedroom lights to off.
Changed status of the kitchen lights to off.
Changed status of the garage lights to off.
Request complete.
Fulfilling request: Turn off the lights in all rooms and play a movie
in which Tom Cruise is a lawyer in the living room.
Changed status of the living room lights to off.
Changed status of the bedroom lights to off.
```

```
Changed status of the kitchen lights to off.
Changed status of the garage lights to off.
Playing A Few Good Men on the TV in the living room.
Request complete.
```

The planner executed each request flawlessly, and you didn't have to write any code. When the user asks something such as "*turn off the lights in all rooms*," the planner realizes that it needs to call the function for the kitchen, the bedroom, the living room, and the garage.

When the user asks for a movie with Tom Cruise as a lawyer, the planner realizes that it needs to call a semantic function to find the name of the movie before calling the `OperateTV` function to put the movie on the TV, again without you having to write code for this explicitly.

Summary

In this chapter, we introduced the planner, a powerful function that allows users to execute very complex workflows with minimal effort on the developer's part. We learned when to use the planner and what the potential issues are. We also learned how to use the planner, as well as how to write descriptions for the functions in our plugins in a way that makes it easier for the planner to combine them. We then saw a longer example of how to use the planner to let a user combine native and semantic functions.

In the next chapter, we will explore ways to make external data available to Semantic Kernel. Later we will pair search with external data to allow models to use large amounts of data that exceed models' context windows.

6

Adding Memories to Your AI Application

In the previous chapter, we learned how to use planners to give our users the ability to ask our application to perform actions that we did not program explicitly. In this chapter, we are going to learn how to use external data, so that we can bring recent information and keep information between user sessions. For now, we are going to use small amounts of external data that our users may have given us by saving it to **memory**. Learning how to use memory will enable us to greatly expand the capabilities of AI models.

This is a building block for the next chapter, in which we are going to learn techniques to use amounts of data that vastly exceed the context window of existing models. As you may remember, a *context window* is the maximum size of the input you can send to an AI service. By using memory, you can save a large amount of data and only send portions of the data in each call.

We will start by understanding how LLMs convert words into meaning by using **embeddings** and then compare phrases with similar meanings to recall data from memory. Later in the chapter, we will see how to keep historical data in a chat application.

In this chapter, we'll be covering the following topics :

- Creating embeddings for text data
- Storing data in memory and recovering it to use in your prompts
- Using a plugin to keep track of a chat that your user is having with your application
- Using summarization to keep track of long chats

By the end of the chapter, you will have learned how to help your application remember information entered by the user and retrieve it when needed.

Technical requirements

To complete this chapter, you will need to have a recent, supported version of your preferred Python or C# development environment:

- For Python, the minimum supported version is Python 3.10, and the recommended version is Python 3.11

- For C#, the minimum supported version is .NET 8

In this chapter, we will call OpenAI services. Given the amount that companies spend on training these LLMs, it's no surprise that using these services is not free. You will need an **OpenAI API** key, either directly through **OpenAI** or **Microsoft**, via the **Azure OpenAI** service.

If you are using .NET, the code for this chapter is at `https://github.com/PacktPublishing/Building-AI-Applications-with-Microsoft-Semantic-Kernel/tree/main/dotnet/ch6`.

If you are using Python, the code for this chapter is at `https://github.com/PacktPublishing/Building-AI-Applications-with-Microsoft-Semantic-Kernel/tree/main/python/ch6`.

You can install the required packages by going to the GitHub repository and using the following: `pip install -r requirements.txt`.

Defining memory and embeddings

LLMs provided by AI services such as OpenAI are **stateless**, meaning they don't retain any memory of previous interactions. When you submit a request, the request itself contains all the information the model will use to respond. Any previous requests you submitted have already been forgotten by the model. While this stateless nature allows for many useful applications, some situations require the model to consider more context across multiple requests.

Despite their immense computing power, most LLMs can only work with small amounts of text, about one page at a time, although this has been increasing recently — the new GPT-4 Turbo, released in November 2023, can receive 128,000 tokens as input, which is about 200 pages of text. Sometimes, however, there are applications that require a model to consider more than 200 pages of text — for example, a model that answers questions about a large collection of academic papers.

Memories are a powerful way to help Semantic Kernel work by providing more context for your requests. We add memory to Semantic Kernel by using a concept called **semantic memory search**, where textual information is represented by vectors of numbers called **embeddings**. Since the inception of computing, text has been converted into numbers to help computers compare different texts. For example, the ASCII table that converts letters into numbers was first published in 1963. LLMs convert much more than a single character at a time, using embeddings.

Embeddings take words and phrases as inputs and output a long list of numbers to represent them. The length of the list varies depending on the embedding model. Importantly, words and phrases with similar meanings are close to each other in a numerical sense; if one were to calculate a distance between the numeric components of the embeddings of two similar phrases, it would be smaller than the distance between two sentences with very different meanings.

We will see a complete example in the *Embeddings in action* section, but for a quick example, the difference between the one-word phrases "*queen*" and "*king*" is much smaller than the difference between the one-word phrases "*camera*" and "*dog.*"

Let's take a deeper look.

How does semantic memory work?

Embeddings are a way of representing words or other data as vectors in a high-dimensional space. Embeddings are useful for AI models because they can capture the meaning and context of words or data in a way that computers can process. An embedding model takes a sentence, paragraph, or even some pages of text and outputs the corresponding embedding vector:

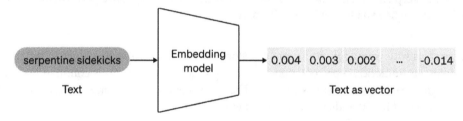

Figure 6.1 – Embedding model

The current highest-performing embedding model that OpenAI makes available to users is called `text-embedding-3-large` and can convert an input of up to 8,191 tokens (about 12 pages of text) into a vector of 3,072 dimensions represented by real numbers.

OpenAI also makes additional embedding models available at different prices and performance points, such as `text-embedding-3-small` and `text-embedding-ada-2`. At the time of writing, the `text-embedding-3-small` model offers better performance than the `text-embedding-ada-2` model, and it's five times cheaper.

You, as a developer, can store text data (including user requests and responses provided by the AI service) as embedding vectors. It's important to know that this will not necessarily make the data smaller. For a given embedding model, embedding vectors will always be the same length. For example, for the `text-embedding-3-large` model, the embedding vector length is always 3,072. If you store the word "*No*" using this model, it will use a vector of 3,072 real numbers that would take 12,228 bytes of memory, a lot more than the string "No", which can usually be stored in two bytes. On the other hand, if you embed 12 pages of text, their embedding vector length will also be 3,072 and take 12,228 bytes of memory.

You can use embeddings to recall context that has been given to your application long before a request is made. For example, you could store all the chats you have had with a user in a database. If the user told you months ago that their favorite city is Paris in France, this information can be saved. Later, when the user asks what the biggest attraction is in the city they like the most, you can search for their favorite city in the database you created.

How are vector databases different from KernelContext?

In previous chapters, we used variables of the type `KernelContext` to pass information to functions. A `KernelContext` variable can be serialized to disk, and therefore, you could use it to store and remember things that your application has already been told.

The difference is that a `KernelContext` variable is a collection of key/value pairs. For each piece of information you store, you have to provide a key, and later, you have to use the same key to retrieve it. Vector databases, on the other hand, retrieve information by similarity, so you can retrieve a piece of information even if you don't know the key used to store it.

Another difference is that, if you want, a vector database can return just the subset of information that is similar to the information you requested, while if you have a `KernelContext` variable, you need to keep all information available all the time, which may cause performance and capacity issues when you have a lot of information.

As the user chats with your application, you can store each command the user types in the chat as an embedding, its numerical representation. Then, when the user enters a new command, you can search through everything the user has ever typed before by comparing the embeddings of what the user just entered to things that they have entered before.

Because the application is using embedding representations that encode meaning, the user may have said "*my favorite city is Paris*" months ago and may ask "*what's the biggest attraction in the city I like the most*" now. A string search would not find a match between "*city I like the most*" and "*favorite city*," but these two sentences will have embedding vectors that are close to each other, and a semantic search would return "*favorite city*" as a close match for "*city I like the most*." In this case, a close match is exactly what you need.

Let's see how to create embeddings with an example.

Embeddings in action

This subsection only has Python code, because OpenAI does not provide a C# API and we would need to call the REST API directly. This subsection will show embedding values to help you understand the embedding concepts, but you don't need to implement the code to understand it.

To start, we will need to import some libraries. In Python, linear algebra calculations are in the numpy library, so we need to import that:

```
from openai import OpenAI
from typing import Tuple
```

```
import numpy as np
from numpy.linalg import norm
import os
```

Here, we will generate embeddings for three sentences and compare them to one another.

First, we will write a function (`get_embedding`) that generates the embeddings:

```
def get_embedding(text: str) -> Tuple[float]:
    client = OpenAI(api_key=os.getenv("OPENAI_API_KEY"))
    response = client.embeddings.create(
        input=text,
        model="text-embedding-3-small"
    )
    return response.data[0].embedding
```

The preceding function is a straightforward function call, simply instantiating a connection to OpenAI and calling the `embeddings.create` method, using the `text-embedding-3-small` model.

Then, to compare how similar one embedding is to another, we will use a **cosine similarity** measure. This is a very simple and widely used measure to compare how close multidimensional vectors are to one another. When multidimensional vectors representing meanings (such as embedding vectors) are close to one another, they have similar meanings. The value returned can be between `0.0` and `1.0`, with `1.0` meaning they are very similar.

```
def similarity(A: np.array, B: np.array) -> float:
    # compute cosine similarity
    cosine = np.dot(A,B)/(norm(A)*norm(B))
    return cosine
```

The last step is to call the functions with the phrases we want to check:

```
if __name__ == "__main__":
    load_dotenv()
    king = get_embedding("The king has been crowned")
    queen = get_embedding("The queen has been crowned")
    linkedin = get_embedding("LinkedIn is a social media platform for
professionals")

    print(similarity(king, queen))
    print(similarity(king, linkedin))
    print(similarity(queen, linkedin))
```

The three phrases are `"The king has been crowned"`, `"The queen has been crowned"`, and `"LinkedIn is a social media platform for professionals"`. We expect the first and second phrases to be similar, and both to be different from the third phrase.

As expected, this is what we get, remembering that the numbers go between 0.0 and 1.0, with 0.0 meaning dissimilar and 1.0 meaning perfect match:

```
0.8684853246664367
0.028215574794606412
0.046607036099519175
```

If you want to see the embeddings themselves, you can print them, remembering that they use 1,536 real numbers for the `text-embedding-3-small` model.

The following code prints the first 10 embedding values:

```
for i in range(0, 10):
    print(king[i])
```

Here's the result:

```
-0.009829566814005375
-0.00965618155896637
0.024287164211273193
0.01408415473997593
-0.03662413731217384
-0.004041192218037605
-0.00032176158856600523
0.046813808381557465
-0.03235621005296707
-0.04099876061081886
```

Now that we understand a little more about how embeddings work, let's see how to use them with LLMs.

Using memory within chats and LLMs

As we have seen before, models have a size limit called a context window. The size limit includes both the prompt with the user request and the response. The default context window for a model such as GPT-3.5, for example, is 4,096 bytes, meaning that both your prompt, including the user request, and the answer that GPT-3.5 provides can have at most 4,096 bytes; otherwise, you will get an error, or the response will cut off in the middle.

If your application uses a lot of text data, for example, a 10,000-page operating manual, or allows people to search and ask questions about a database of hundreds of documents with each one having 50 pages, you need to find a way of including just the relevant portion of this large dataset with your prompt. Otherwise, the prompt alone could be larger than the context window, resulting in an error, or the remaining context window could be so short that there would be no space for the model to provide a good answer.

One way in which you could work around this problem is by summarizing each page into a shorter paragraph and then generating an embedding vector for each summary. Instead of including all the pages in your prompt, you can use something such as cosine similarity to search for the relevant pages by comparing their embeddings with the request embeddings, and then include only the summaries of the relevant pages in the prompt, saving a lot of space.

Another reason to use memory is to keep data between sessions, or even between prompts. For example, as we suggested in the *How does semantic memory work?* section, your user may have told you that their favorite city is Paris, and when they ask for a guide for their favorite city, you don't need to ask again; you just need to search for their favorite city.

To find the data in the memory that is relevant to our prompt, we could use something such as the cosine distance shown in the previous section. In practice, the Semantic Kernel SDK already provides you with a search function, so you don't need to implement it yourself. In addition, you can use several third-party vector databases, each with its own search functions.

Here's a list of all databases that you can use out of the box:

Database name	Python	C#
Azure Cosmos DB for MongoDB vCore		✓
Azure AI Search	✓	✓
Azure PostgreSQL Server	✓	
Chroma	✓	✓
DuckDB	✓	
Milvus		✓
MongoDB Atlas Vector Search	✓	✓
Pinecone	✓	✓
PostgreSQL	✓	✓
Qdrant	✓	
Redis	✓	
SQLite	✓	
Weaviate	✓	✓

Table 6.1 — Vector databases and compatibility with Semantic Kernel

In addition to the databases listed, there's another one, called `VolatileMemoryStore`, which represents the RAM of the machine you're running your code on. That database is not persistent, and its contents are discarded when the code finishes running, but it's fast and free and can be easily used during development.

Using memory with Microsoft Semantic Kernel

In the following example, we will store some information about the user and then use the `TextMemorySkill` core skill to retrieve it directly inside a prompt. Core skills are skills that come out of the box with Semantic Kernel. `TextMemorySkill` has functions to put text into memory and retrieve it.

In the following example, our use case will be of a user who tells us their favorite city and favorite activity. We will save those to memory and then we will retrieve them and provide an itinerary based on the saved information.

We start by importing the libraries that we usually import, plus a few memory libraries that will be described later.

Python

```python
import asyncio
import semantic_kernel as sk
from semantic_kernel.connectors.ai.open_ai import OpenAITextEmbedding,
OpenAIChatCompletion, OpenAIChatPromptExecutionSettings

from semantic_kernel.functions import KernelArguments, KernelFunction
from semantic_kernel.prompt_template import PromptTemplateConfig
from semantic_kernel.utils.settings import openai_settings_from_dot_
env

from semantic_kernel.memory.volatile_memory_store import
VolatileMemoryStore
from semantic_kernel.memory.semantic_text_memory import
SemanticTextMemory
from semantic_kernel.core_plugins.text_memory_plugin import
TextMemoryPlugin
```

C#

```csharp
using Microsoft.SemanticKernel;
using Microsoft.SemanticKernel.Memory;
using Microsoft.SemanticKernel.Plugins.Memory;
using Microsoft.SemanticKernel.Connectors.OpenAI;
#pragma warning disable SKEXP0003, SKEXP0011, SKEXP0052
```

Note that the memory functions in Python are asynchronous, so we must include the `asyncio` library. Also, at the time of writing, the memory functions in C# are marked as experimental, so you have to disable the experimental warnings with the `#pragma` command.

Now, let's create a kernel:

Python

```
def create_kernel() -> tuple[sk.Kernel, OpenAITextEmbedding]:
    api_key, org_id =  openai_settings_from_dot_env()
    kernel = sk.Kernel()
    gpt = OpenAIChatCompletion(ai_model_id="gpt-4-turbo-preview", api_
key=api_key, org_id=org_id, service_id="gpt4")
    emb = OpenAITextEmbedding(ai_model_id="text-embedding-ada-002",
api_key=api_key, org_id=org_id, service_id="emb")
    kernel.add_service(emb)
    kernel.add_service(gpt)
    return kernel, emb

async def main():
    kernel, emb = create_kernel()
    memory = SemanticTextMemory(storage=VolatileMemoryStore(),
embeddings_generator=emb)
    kernel.add_plugin(TextMemoryPlugin(memory), "TextMemoryPlugin")
```

C#

```
var builder = Kernel.CreateBuilder();
builder.AddOpenAIChatCompletion("gpt-4-turbo-preview", apiKey, orgId);
var kernel = builder.Build();

var memoryBuilder = new MemoryBuilder();
memoryBuilder.WithMemoryStore(new VolatileMemoryStore());
memoryBuilder.WithOpenAITextEmbeddingGeneration("text-embedding-3-
small", apiKey);
var memory = memoryBuilder.Build();
```

Note that we added three items to our kernel:

- An embedding model, which will help us load things into memory. For C#, we can use `text-embedding-3-small`, but at the time of writing, even though Python can use `text-embedding-3-small` as we did in the previous section, the core Python plugins only work with model `text-embedding-ada-002`.

- Memory storage; in this case, `VolatileMemoryStore`, which just stores data temporarily in your computer's RAM.

- A GPT model to generate the itinerary; we're using GPT-4

Also, note that in C#, the memory and the kernel are built in separate commands, while in Python, they are all built together.

Now, let's create a function that adds data to the memory:

Python

```
async def add_to_memory(memory: SemanticTextMemory, id: str, text:
str):
    await memory.save_information(collection="generic", id=id,
text=text)
```

C#

```
const string MemoryCollectionName = "default";
await memory.SaveInformationAsync(MemoryCollectionName, id: "1", text:
"My favorite city is Paris");
await memory.SaveInformationAsync(MemoryCollectionName, id: "2", text:
"My favorite activity is visiting museums");
```

The function to add data to memory simply calls `memory.save_information` in Python and `memory.SaveInformationAsync` in C#. You can keep different groups of information separate by using collections, but in our simple case, we're just going to use `"generic"` for Python and `"default"` for C#, as those are the default collections for the plugins. The `id` parameter does not have to mean anything, but it must be unique by item. If you save multiple items using the same `id` parameter, the last saved item will overwrite the previous ones. It's common to generate GUIDs to ensure some level of uniqueness, but if you are just going to add a few items manually, you can manually ensure that the ids are different.

We can now create a function that generates a travel itinerary:

Python

```
async def tour(kernel: sk.Kernel) -> KernelFunction:
    prompt = """
    Information about me, from previous conversations:
    - {{$city}} {{recall $city}}
    - {{$activity}} {{recall $activity}}
    """.strip()

    execution_settings = kernel.get_service("gpt4").instantiate_
prompt_execution_settings(service_id="gpt4")
    execution_settings.max_tokens = 4000
    execution_settings.temperature = 0.8

    prompt_template_config = PromptTemplateConfig(template=prompt,
execution_settings=execution_settings)
    chat_func = kernel.add_function(
```

```
        function_name="chat_with_memory",
        plugin_name="TextMemoryPlugin",
        prompt_template_config=prompt_template_config,
    )

    return chat_func
```

C#

```csharp
kernel.ImportPluginFromObject(new TextMemoryPlugin(memory));
const string prompt = @"
Information about me, from previous conversations:
- {{$city}} {{recall $city}}
- {{$activity}} {{recall $activity}}

Generate a personalized tour of activities for me to do when I have a
free day in my favorite city. I just want to do my favorite activity.
";

var f = kernel.CreateFunctionFromPrompt(prompt, new
OpenAIPromptExecutionSettings { MaxTokens = 2000, Temperature = 0.8
});
var context = new KernelArguments();
context["city"] = "What is my favorite city?";
context["activity"] = "What is my favorite activity?";


context[TextMemoryPlugin.CollectionParam] = MemoryCollectionName;
```

TextMemoryPlugin gives the ability to use {{recall $question}} to retrieve the contents of the memory inside a prompt without you needing to write any code.

For example, assume that we loaded My favorite city is Paris in our memory. When we load the $city variable with "What's my favorite city" and write {{$city}} {{recall $city}} inside the prompt, Semantic Kernel will replace that line with "What's my favorite city? My favorite city is Paris" inside the prompt.

> **Storing data in memory**
>
> Note that we didn't use a meaningful key name when storing the data in memory (we used "1" and "2"). You also don't need to classify the information before storing it. Some applications simply store everything as they comes, while others use a semantic function to ask Semantic Kernel whether the user input contains personalization information and store it in cases where it does.

Now, let's load the memory and call the prompt:

Python

```
    await add_to_memory(memory, id="1", text="My favorite city is
Paris")
    await add_to_memory(memory, id="2", text="My favorite activity is
visiting museums")

    f = await tour(kernel)

    args = KernelArguments()
    args["city"] = "My favorite city is Paris"
    args["activity"] = "My favorite activity is visiting museums"
    answer = await kernel.invoke(f, arguments=args)
    print(answer)

if __name__ == "__main__":
    asyncio.run(main())
```

C#

```
await memory.SaveInformationAsync(MemoryCollectionName, id: "1", text:
"My favorite city is Paris");
await memory.SaveInformationAsync(MemoryCollectionName, id: "2", text:
"My favorite activity is visiting museums");
var result = await f.InvokeAsync(kernel, context);
Console.WriteLine(result);
```

In the code, we load the information to memory using the add_to_memory function and immediately call our semantic function f. If you are using any memory store other than VolatileMemoryStore, you don't need to implement these two steps in the same session. We will see an example of persisting memory in our **RAG (retrieval-augmented generation)** example in *Chapter 7*.

Results

Note that the model recalled that the user's favorite city is Paris and that their favorite activity is going to museums:

```
Given your love for Paris and visiting museums, here's a personalized
itinerary for a fulfilling day exploring some of the most iconic and
enriching museums in the City of Light:

**Morning: Musée du Louvre**
```

```
Start your day at the Louvre, the world's largest art museum and
a historic monument in Paris. Home to thousands of works of art,
including the Mona Lisa and the Venus de Milo, the Louvre offers an
unparalleled experience for art lovers. Arrive early to beat the
crowds and spend your morning marveling at the masterpieces from
across the world. Don't forget to walk through the Tuileries Garden
nearby for a peaceful stroll.

**Afternoon: Musée d'Orsay**

Next, head to the Musée d'Orsay, located on the left bank of the
Seine. Housed in the former Gare d'Orsay, a Beaux-Arts railway
station, the museum holds the largest collection of Impressionist and
Post-Impressionist masterpieces in the world. Spend your afternoon
admiring works by Monet, Van Gogh, Renoir, and many others.

**Late Afternoon: Musée de l'Orangerie**

Conclude your day of museum visits at the Musée de l'Orangerie,
located in the corner of the Tuileries Gardens. This gallery is famous
for housing eight large Water Lilies murals by Claude Monet, displayed
in two oval rooms offering a breathtaking panorama of Monet's garden-
inspired masterpieces. The museum also contains works by Cézanne,
Matisse, Picasso, and Rousseau, among others.

**Evening: Seine River Walk and Dinner**

After an enriching day of art, take a leisurely walk along the Seine
River. The riverside offers a picturesque view of Paris as the city
lights begin to sparkle. For dinner, choose one of the numerous
bistros or restaurants along the river or in the nearby neighborhoods
to enjoy classic French cuisine, reflecting on the beautiful artworks
and memories created throughout the day.
```

Note that if you pay for a subscription to any of the vector database providers listed in *Table 6.1*, you can simply replace the `VolatileMemoryStore` constructor with their constructor; for example, if you are using Pinecone, you will use `Pinecone(apiKey)`, and the memory will be persisted in that database and available to the user the next time they run your application. We will see an example with Azure AI Search in *Chapter 7*.

Now, let's see how we can use memory in a chat with a user.

Using memory in chats

Memory is typically used in chat-based applications. All the applications that we built in earlier chapters were *one-shot* — all the information required to complete a task is part of the request submitted by the user plus whatever modifications we make to the prompt in our own code, for example, by including the user-submitted prompt inside a variable in `skprompt.txt`, or modifying their prompt using string manipulation. All questions and answers that happened before are ignored. We say that the AI service is *stateless*.

Sometimes, however, we want the AI service to remember requests that have been made before. For example, if I ask the app about the largest city in India by population, the application will respond that it is *Mumbai*. If I then ask "*how is the temperature there in the summer*," I would expect the application to realize that I'm asking about the temperature in Mumbai, even though my second prompt does not include the name of the city.

As we mentioned in *Chapter 1*, the brute-force solution is to simply repeat the whole history of the chat with every new request. Therefore, when the second request is submitted by the user, we could silently attach their first request and the response our AI service provided to it and then submit everything together again to the AI service.

Let's see how to do this next. We start with the usual imports:

Python

```
import asyncio
import semantic_kernel as sk
from semantic_kernel.connectors.ai.open_ai import
OpenAIChatCompletion, OpenAIChatPromptExecutionSettings
from semantic_kernel.functions import KernelFunction
from semantic_kernel.prompt_template import PromptTemplateConfig,
InputVariable
from semantic_kernel.core_plugins import ConversationSummaryPlugin
from semantic_kernel.contents.chat_history import ChatHistory
from semantic_kernel.utils.settings import openai_settings_from_dot_
env
```

C#

```
using Microsoft.SemanticKernel;
using Microsoft.SemanticKernel.Connectors.OpenAI;
using Microsoft.SemanticKernel.Plugins.Core;
#pragma warning disable SKEXP0003, SKEXP0011, SKEXP0052, SKEXP0050
```

Note that in C#, since several components of the Semantic Kernel package are still in pre-release, you need to disable the experimental warnings using a `#pragma` directive.

After importing the libraries, we create the kernel:

Python

```
def create_kernel() -> sk.Kernel:
    api_key, org_id = openai_settings_from_dot_env()
    kernel = sk.Kernel()
    gpt = OpenAIChatCompletion(ai_model_id="gpt-4-turbo-preview", api_
key=api_key, org_id=org_id, service_id="gpt4")
```

```python
    kernel.add_service(gpt)

    # The following execution settings are used for the
ConversationSummaryPlugin
    execution_settings = OpenAIChatPromptExecutionSettings(
        service_id="gpt4", max_tokens=ConversationSummaryPlugin._max_
tokens, temperature=0.1, top_p=0.5)

    prompt_template_config = PromptTemplateConfig(
        template=ConversationSummaryPlugin._summarize_conversation_
prompt_template,
        description="Given a section of a conversation transcript,
summarize it",
        execution_settings=execution_settings,
    )

    # Import the ConversationSummaryPlugin
    kernel.add_plugin(
        ConversationSummaryPlugin(kernel=kernel, prompt_template_
config=prompt_template_config),
        plugin_name="ConversationSummaryPlugin",
    )

    return kernel
```

C#

```csharp
var (apiKey, orgId) = Settings.LoadFromFile();
var builder = Kernel.CreateBuilder();
builder.AddOpenAIChatCompletion("gpt-4-turbo-preview", apiKey, orgId);
var kernel = builder.Build();
kernel.ImportPluginFromObject(new ConversationSummaryPlugin());
```

Our kernel just needs a chat completion service. I'm using GPT-4, but GPT-3.5 also works. I am also adding ConversationSummaryPlugin, which will be used in the last subsection of this chapter, *Reducing history size with summarization*. We will explain it in detail later, but as the name implies, it summarizes conversations.

Now, let's create the main chat function:

Python

```python
async def create_chat_function(kernel: sk.Kernel) -> KernelFunction:
    # Create the prompt
    prompt = """
    User: {{$request}}
```

```
    Assistant:  """

    # These execution settings are tied to the chat function, created
below.
    execution_settings = kernel.get_service("gpt4").instantiate_
prompt_execution_settings(service_id="gpt4")
    chat_prompt_template_config = PromptTemplateConfig(
        template=prompt,
        description="Chat with the assistant",
        execution_settings=execution_settings,
        input_variables=[
            InputVariable(name="request", description="The user
input", is_required=True),
            InputVariable(name="history", description="The history of
the conversation", is_required=True),
        ],
    )

    # Create the function
    chat_function = kernel.add_function(
        prompt=prompt,
        plugin_name="Summarize_Conversation",
        function_name="Chat",
        description="Chat with the assistant",
        prompt_template_config=chat_prompt_template_config,)

    return chat_function
```

C#

```
const string prompt = @"
Chat history:
{{$history}}

User: {{$userInput}}
Assistant:";

var executionSettings = new OpenAIPromptExecutionSettings {MaxTokens =
2000,Temperature = 0.8,};

var chatFunction = kernel.CreateFunctionFromPrompt(prompt,
executionSettings);
var history = "";
var arguments = new KernelArguments();
arguments["history"] = history;
```

Now, let's write the main loop of our program:

Python

```python
async def main():
    kernel = create_kernel()
    history = ChatHistory()

    chat_function = await create_chat_function(kernel)

    while True:
        try:
            request = input("User:> ")
        except KeyboardInterrupt:
            print("\n\nExiting chat...")
            return False
        except EOFError:
            print("\n\nExiting chat...")
            return False

        if request == "exit":
            print("\n\nExiting chat...")
            return False

        result = await kernel.invoke(
            chat_function,
            request=request,
            history=history,
        )

        # Add the request to the history
        history.add_user_message(request)
        history.add_assistant_message(str(result))

        print(f"Assistant:> {result}")

if __name__ == "__main__":
    asyncio.run(main())
```

C#

```
var chatting = true;
while (chatting) {

    Console.Write("User: ");
    var input = Console.ReadLine();

    if (input == null) {break;}
    input = input.Trim();

    if (input == "exit") {break;}

    arguments["userInput"] = input;
    var answer = await chatFunction.InvokeAsync(kernel, arguments);
    var result = $"\nUser: {input}\nAssistant: {answer}\n";

    history += result;
    arguments["history"] = history;

    // Show the bot response
    Console.WriteLine(result);
}
```

The main loop of our program runs until the user enters the word `"exit"`. Otherwise, we submit the user request to the AI service, collect its answer, and add both to the `history` variable, which we also submit as part of our request.

Although this solves the problem of always having the whole history, it becomes prohibitively expensive as prompts start to get larger and larger. When the user submits their request number N, the `history` variable contains their requests 1, ..., $N-1$, and the chatbot answers 1, ..., $N-1$ along with it. For large N, in addition to being expensive, this can exceed the context window of the AI service and you will get an error.

The solution is to only pass a summary of the history to the AI service. It's fortunate that summarizing conversations is something that even older models can do very well. Let's see how to easily do it with Semantic Kernel.

Reducing history size with summarization

If you want to reduce the prompt without losing much context, you can use the AI service to summarize what has already happened in the conversation. To do so, you can use the `SummarizeConversation` function of `ConversationSummaryPlugin` that we imported when creating the kernel. Now, instead of repeating the whole history in the prompt, the summary will have up to 1,000 tokens regardless of the conversation size, which should be plenty for most use cases. To summarize the history in the `$history` variable, simply call `{{ConversationSummaryPlugin.SummarizeConversation $history}}` in your prompt.

It is still possible to lose details after too much summarization. If you try to summarize 1,000 pages in 1,000 words, something will be lost. To prevent this problem, most applications have limits on how long conversations can go. For example, at the time of writing, Microsoft Copilot conversations have a limit of 20 interactions, and you must restart the conversation (with an empty memory) after that.

The change to the code is shown as follows; you just need to change the contents of the `prompt` variable. The change will add a conversation summary to the prompt, which will remind the LLM of everything that went on before. The summary will not be displayed to the user.

Python

```
prompt = """
Chat history:
{{ConversationSummaryPlugin.SummarizeConversation $history}}
User: {{$userInput}}
Assistant:
"""
```

C#

```
const string prompt = @"
Chat history:
{{ConversationSummaryPlugin.SummarizeConversation $history}}

User: {{$userInput}}
ChatBot:";
```

Results

```
User:> What is the largest city by population in Western Europe?
Assistant:> The largest city by population in Western Europe is
London, United Kingdom.
```

Note that in the next question, I'll use the word `there` to mean London. Since the history summary is included as part of the conversation, even though my next question doesn't explicitly name London, the prompt that goes to the AI contains that information:

```
User:> Are there any famous people who lived there?
Assistant:> Yes, London has been home to many famous people throughout
history. Some notable individuals include:
```

```
1. **William Shakespeare** - The renowned playwright and poet lived in
London for most of his career.
2. **Charles Dickens** - The famous novelist, known for works like
"Oliver Twist" and "A Christmas Carol," spent much of his life in
London.
3. **Winston Churchill** - The iconic British Prime Minister during
World War II was born in Woodstock but lived and worked in London for
much of his life.
```

Note that the answer to the previous question was correct. Shakespeare, Dickens, and Churchill all lived in London. Now, I'll refer to Shakespeare just by its position on the list and to London simply as `that city`, and because we're keeping track of history, the kernel will know that I mean Shakespeare and London:

```
User:> What is a famous play from the first one set in that city?
Assistant:> A famous play from William Shakespeare that is set in
London is "Henry V." This historical play, part of Shakespeare's
series on the English monarchs, includes scenes that are set in
London, notably the English court. It portrays the events before and
after the Battle of Agincourt during the Hundred Years' War, with
significant portions reflecting on the life and times in London during
the early 15th century.
```

Again, the AI gets the answer correct. The `Henry V` play is actually set in London.

Let's exit the chat:

```
User:> exit|
Exiting chat...
```

Summary

In this chapter, we learned how to add and retrieve information from memory, and how to easily include the memory in your prompts. LLMs are stateless and limited by their prompt sizes, and in this chapter, we learned techniques to save information between sessions and reduce prompt sizes while still including relevant portions of the conversation in the prompt.

In the next chapter, we will see how to use a vector database to retrieve a lot more information from memory and use a technique called **retrieval-augmented generation** (**RAG**) to organize and present that information in a useful way. This technique is often used in enterprise applications, as you trade off a little bit of the creativity offered by LLMs, but get back additional precision, the ability to show references, and the ability to use a lot of data that you own and have control over.

For our application, we are going to load thousands of academic articles into a vector database and have Semantic Kernel search for a topic and summarize the research for us.

Part 3:
Real-World Use Cases

In this part, we see how Semantic Kernel can be used in real-world problems. We learn how using the retrieval-augmented generation (RAG) technique can allow AI models to use large amounts of data, including very recent data that was not available when the AI service was trained. We conclude by learning how to use ChatGPT to distribute an application we wrote to hundreds of millions of users.

This part includes the following chapters:

- *Chapter 7, Real-World Use Case – Retrieval-Augmented Generation*
- *Chapter 8, Real-World Use Case – Making Your Application Available on ChatGPT*

7

Real-World Use Case – Retrieval-Augmented Generation

In the previous chapter, we learned how to augment our kernel with memories, which enables our applications to be much more personalized. Cloud-based AI models, such as OpenAI's GPT, usually have knowledge cut-offs that are a few months old. They also usually don't have domain-specific knowledge, such as the user manuals of the products your company makes, and don't know the preferences of your users, such as their favorite programming language or their favorite city. The previous chapter taught you ways to augment the knowledge of models by keeping small pieces of knowledge in memory and retrieving them as needed.

In this chapter, we're going to show you how to expand the data that's available to your AI application. Instead of using a small amount of data that fits in the prompt, we're going to use a large amount of data with a **retrieval-augmented generation** (**RAG**) application that combines the latest generative AI models with recent specialized information to answer questions about a specific topic – in our case, academic articles about AI.

RAG takes advantage of the fact that lots of institutions have useful data that wasn't part of the data that was used to train OpenAI's GPT. This gives these institutions a way of putting this data to use while still taking advantage of the generative power of GPT.

In this chapter, we'll be covering the following topics:

- Creating a document index with the Azure AI Search service

- Loading a large number of documents to the index

- Creating an application that searches the index and uses AI to write an answer based on the data it found

By the end of this chapter, you will have created an application that uses a large amount of recent data and uses AI to find and combine the data in a user-friendly way.

Technical requirements

To complete this chapter, you will need to have a recent, supported version of your preferred Python or C# development environment:

- For Python, the minimum supported version is Python 3.10, and the recommended version is Python 3.11

- For C#, the minimum supported version is .NET 8

In this chapter, we will call OpenAI services. Given the amount that companies spend on training these LLMs, it's no surprise that using these services is not free. You will need an **OpenAI API** key, obtained either directly through **OpenAI** or **Microsoft**, via the **Azure OpenAI** service.

If you are using .NET, the code for this chapter is at `https://github.com/PacktPublishing/Building-AI-Applications-with-Microsoft-Semantic-Kernel/tree/main/dotnet/ch7`.

If you are using Python, the code for this chapter is at `https://github.com/PacktPublishing/Building-AI-Applications-with-Microsoft-Semantic-Kernel/tree/main/python/ch7`.

To create a document index, you will need a free trial of Microsoft Azure AI Search.

You can install the required packages by going to the GitHub repository and using the following: `pip install -r requirements.txt`.

Why would you need to customize GPT models?

GPT models are already very useful without any customizations. When your user types a request, you, as a programmer, could simply forward the request to the GPT model (such as GPT-3.5 or GPT-4), and, in many cases, the unaltered response from the model is good enough. However, in many cases, the responses aren't good enough. There are three categories of problems with responses:

- **Non-text functionality**: In some cases, the response you want is not text-based. For example, you may want to allow your user to turn a light on or off, perform complex math, or insert records into a database.

- **Lack of context**: Models can't accurately answer questions if they haven't been exposed to the data that contains the answer. Despite being trained with immense amounts of data, there's a lot of data that LLMs haven't been exposed to. At the time of writing, the cut-off date for data used to train GPT 3.5 and GPT-4 is September 2021, although there is a preview version of GPT-4 called GPT-4 Turbo with a cut-off date of December 2023 (you can see the cut-off dates of models at `https://platform.openai.com/docs/models/`.) In addition, models don't have access to proprietary data, such as the internal documents of your company.

- **Unsupported formats**: LLMs such as GPT-3.5 Turbo and GPT-4 have been trained to provide answers in text with a conversational tone. In some cases, you might want the answers to be provided in a specific format, such as JSON. Since models haven't been trained for that, they may provide inconsistent answers. It's not uncommon for a low percentage of responses to use an incorrect format, even though your prompt was very specific. For example, you may have added `Answer only with Y or N` to your prompt, but some requests return responses such as `Yes` (instead of `Y`) or `The answer is yes`, which requires adding code to validate the answer.

We showed you how to solve the first issue (non-text functionality) using Semantic Kernel via native functions, as shown in *Chapter 3*. However, if the problem with the responses you're getting is a lack of context or format, you can use the techniques depicted in the following diagram:

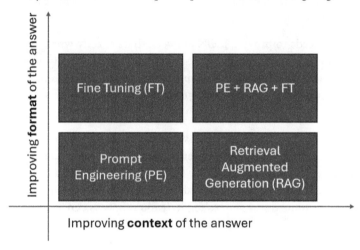

Figure 7.1 – Techniques to improve responses

The first technique you should always try is **prompt engineering**, something we covered in detail in *Chapter 2*. Prompt engineering is easy to do and test: it can be used both to give new data to the LLM (improving context) and to provide some examples of how you want the answer to look (improving format).

For example, let's say you're building an application that gives your team suggestions of places to go for lunch, something that's always a challenge among teams of developers. Instead of simply asking `Where should we go for lunch?`, you will get much better results by adding context and format specifications, such as `We are a team of six developers aged 25-38, two of us are vegetarians, and we are looking for places to have lunch near the Eiffel Tower on a Friday. We want to spend less than 20 euro per person and we don't want to spend more than 90 minutes having lunch. Please provide your answer with the name of the place, their website, their average price, and their street address`. The format specification is the last sentence.

The main downside is that the more data you want to provide and the more complex the instructions, the larger your prompts will become, resulting in additional costs and latency.

Besides providing examples through prompt engineering, another technique you can use to improve the format of your answer is to fine-tune your model. Fine-tuning allows you to provide hundreds or thousands of examples of questions and answers to an existing model (for example, GPT-3.5) and save a new, fine-tuned model.

One example of successful fine-tuning is to show thousands of examples of the way you expect JSON output to look. Since you are providing thousands of examples, you can't pass this on to every prompt because the prompt will become too large. You can create a file that contains thousands of questions and JSON answers and use the OpenAI fine-tuning API or fine-tuning UI to create a custom GPT model that has been trained with your additional examples. The result will be a model that is a lot better at providing JSON answers, and worse at everything else.

If your application only needs to provide JSON answers, that's exactly what you need. Microsoft Semantic Kernel does not help with fine-tuning, so techniques for fine-tuning are outside the scope of this book. If you want to learn more about fine-tuning, this online article from Sebastian Raschka, a Packt author, can help: `https://magazine.sebastianraschka.com/p/finetuning-large-language-models`.

In practice, one of the most common problems is that the LLM will not have enough context to provide the answers you want. This can happen even if the data that's required to provide the answer has been used to train the model: since LLMs are trained with a lot of data, you may need to add relevant data to your request to help the model recall the data that's relevant to your request from the large amount of data it was trained with. For example, if you simply ask GPT `Who is the best football player of all time?`, it may not know whether you mean association football (soccer) or NFL (American) football.

In some other cases, as discussed previously when we mentioned the cut-off date and private data examples, the model has never seen the data required to answer the question, and you need to show it to the model as you are making the request.

To an extent, you can solve both problems with prompt engineering:

- You can tell the model to play a role. For example, you can add `you are a Python software engineer` to prime the model to respond more technically, or `you are a five-year-old child` to prime the model to respond more simply.

- You can give the model some data examples. For example, you can add `If the user says 'the earth is flat', reply with 'misinformation'; if the user says 'the moon landing was fake', reply with 'misinformation'; if the user says 'birds are real', reply with "true"` to your prompt, either directly or by using prompt templates in semantic functions.

- You can add some fields to your prompt template and fill them in real time. For example, you can get today's date from the system and create a prompt that states `the difference between $today and July 4th, 1776, in days is…"`, replacing `$today` dynamically, and therefore passing recent information to the model.

The first downside of prompt engineering is that the more data you need to pass, the larger your prompts will get, which will make the prompts more expensive. It will also increase latency as it will take longer for the LLM to process long prompts.

Even if your budget can support the additional cost and your users are extremely patient and don't mind waiting for the answers, there are still two problems. The first is that the accuracy of LLMs decreases [1] as prompts get larger. The second is that at some point, you may run out of space in the context window of the model. For example, let's say you work for a company that manufactures cars, and you want to help a user find the answer to a question about their car in the user manual, but it's 300 pages long. Even if you were to solve all previous problems, you can't pass the whole manual in the prompt because it doesn't fit.

The solution that works best is to break your user manual into several chunks and save these chunks to an index. When the user asks a question, you can use a search algorithm to return the most relevant chunks by using something such as cosine similarity, as shown in *Chapter 6*. Then, you only need to pass the relevant chunks to the prompt. The name of this technique is RAG and it's widely used. Semantic Kernel makes it easy to implement it, but you also need an index. Let's delve into the details.

Retrieval-augmented generation

RAG is an approach that combines the powers of pre-trained language models with information retrieval to generate responses based on a large corpus of documents. This is particularly useful for generating informed responses that rely on external knowledge not contained within the model's training dataset.

RAG involves three steps:

- **Retrieval**: Given an input query (for example, a question or a prompt), you use a system to retrieve relevant documents or passages from your data sources. This is typically done using embeddings.

- **Augmentation**: The retrieved documents are then used to augment the input prompt. Usually, this means creating a prompt that incorporates the data from the retrieval step and adds some prompt engineering.

- **Generation**: The augmented prompt is then fed into a generative model, usually GPT, which generates the output. Because the prompt contains relevant information from the retrieved documents, the model can generate responses that are informed by that external knowledge.

In addition to providing additional and more recent information to an AI service, RAG can help with **grounding**. Grounding is the process of tying the language model's responses to accurate, reliable, and contextually appropriate knowledge or data. This can be particularly important in scenarios where factual accuracy and relevance are crucial, such as answering questions about science, history, or current events. Grounding helps ensure that the information provided by the model is not only plausible but also correct and applicable to the real world.

When you use RAG, you give the LLM the data that you want it to use to generate your responses. If your data is accurate, reliable, and contextually appropriate, the text that's generated by the LLM using this data has a very high likelihood of also being accurate, reliable, and contextually appropriate. You can even ask the generator step to provide links to the documents it used. We will see this in our example.

Let's say you want to summarize the latest discoveries in models with large context windows. First, you need to retrieve information about the latest discoveries by doing a web search or using a database of academic papers.

To implement RAG, you need a few extra components:

- **Document store**: A large collection of documents that the model can search through to find relevant information. This could be anything from a simple text file, a database, or a more sophisticated document store such as a vector database. In our example, we will use a vector database, **Azure AI Search**, but lots of implementations use out-of-the-box Python components such as `numpy`, which have the advantage of being free.
- **Retrieval system**: The software that's used to find the most relevant documents from the document store based on the input query.

Most vector database vendors provide algorithms that work well with their service, and lately, most solutions have been using vector comparisons such as cosine similarity. For example, services such as Pinecone and Azure AI Search provide document and embedding storage and retrieval algorithms.

In our example, we will create an application that allows you to search for and ask questions about AI papers from the ArXiV database. We downloaded the list of ArXiV IDs, authors, titles, and abstracts for all papers in the *Computation and Language* category that were submitted in 2021 and after. This dataset is available in this book's GitHub repository: `https://github.com/PacktPublishing/Microsoft-Semantic-Kernel/blob/b1187f88f46589f14a768e3ee7b89bef733f1263/data/papers/ai_arxiv_202101.json`.

The dataset contains a total of 36,908 scientific articles. The summaries of their contents are in the `abstract` field and contain over 40 million characters, which would require approximately 10 million tokens, something that's too large for even the largest AI models.

We are going to load all this data into an Azure AI Search index. But before we load the articles, we must create the index.

Creating an index

To store and retrieve large amounts of data, we will need to create an index. To do so, you must have an Azure account and must create an Azure AI Search service. Just search for `Azure AI Search` and click **Create**; you will be asked for a name. You will need the endpoint of the service, which you can find in the **Configuration** tab, shown in *Figure 7.2.* or the Azure AI Search service you created. *Figure 7.2* shows the endpoint for the service you created in the **Url** field, marked in green:

Figure 7.2 – Azure AI Search configuration screen

You will also need an admin key, which you can find under the **Keys** tab for your Azure AI Search service.

Creating a service is just the first step: the service is just a place to store one or more indexes, which are the places where we will store the data. Now that we have a service, we need to write code to create the index.

The field names deserve mentioning. Your life will be a lot easier if you can use some standard names – that is, `Id`, `AdditionalMetadata`, `Text`, `Description`, `ExternalSourceName`, `IsReference`, and `Embedding`. The field names should use that specific capitalization. If you use these names, you can easily use the preview version of the Azure AI Search Semantic Kernel connection, which will make your code much smaller. The text you'll use for searching (abstracts, in our case) should be `Text`. In the following code, I'll map these fields to what we need.

So, let's see how to do that in Python. Later, we'll learn how to do this in C#.

Creating the index with Python

Write the following code in a Python script to create an index:

```
from azure.core.credentials import AzureKeyCredential
from azure.search.documents.indexes import SearchIndexClient
```

First, you need to import the `AzureKeyCredential` function to read your admin key and `SearchIndexClient` to create an object that will allow you to interact with the Azure AI Search service.

Next, we will import several classes for the types we will be using in our index:

```
from azure.search.documents.indexes.models import (
    SearchIndex,
    SearchField,
    SearchFieldDataType,
    SimpleField,
    SearchableField,
    VectorSearch,
    HnswAlgorithmConfiguration,
    HnswParameters,
    VectorSearchAlgorithmKind,
    VectorSearchProfile,
    VectorSearchAlgorithmMetric,
)
```

For fields that we want to search using embeddings, we use the `SearchField` type. For other fields, we use the `SimpleField` type if we don't intend to search for content inside of them, and `SearchableField` if we want them to be searchable by string comparisons.

Next, let's create an API client that will add a new index to the index collection with the `SearchIndexClient` class:

```
def main() -> None:

    index_name = os.getenv("ARXIV_SEARCH_INDEX_NAME")
    service_name = os.getenv("ARXIV_SEARCH_SERVICE_NAME")
    service_endpoint = f"https://{service_name}.search.windows.net/"
    admin_key = os.getenv("ARXIV_SEARCH_ADMIN_KEY")
    credential = AzureKeyCredential(admin_key)

    # Create a search index
    index_client = SearchIndexClient(
        endpoint=service_endpoint, credential=credential)

    index_client.delete_index(index_name)
```

When you are in the development phase, it's not uncommon to have to redesign your index by adding or dropping fields, changing the size of the embeddings, and so on. Therefore, we usually drop and recreate the fields in the script. To drop a field in the preceding snippet, we used the `delete_index` method.

The following code specifies the fields and their properties to help describe which fields the index will contain:

```
fields = [
    SimpleField(name="Id", type=SearchFieldDataType.String, key=True,
sortable=True, filterable=True, facetable=True),
    SearchableField(name="AdditionalMetadata",
type=SearchFieldDataType.String),
    SearchableField(name="Text", type=SearchFieldDataType.String),
    SearchableField(name="Description", type=SearchFieldDataType.
String),
    SearchableField(name="ExternalSourceName",
type=SearchFieldDataType.String),
    SimpleField(name="IsReference", type=SearchFieldDataType.Boolean),
    SearchField(name="Embedding", type=SearchFieldDataType.
Collection(SearchFieldDataType.Single),
                searchable=True, vector_search_dimensions=1536,
vector_search_profile_name="myHnswProfile"),
]
```

Here, we are adding the same fields we have in our dataset to the index: `id`, `authors`, `title`, and `abstract`. In addition, we're adding a field called `Embedding`, where we will put the embedding vectors of the articles' abstracts. For that field, we need to specify a vector search algorithm profile and a vector search dimension. The dimension is the size of the embeddings. Since we're using the new `OpenAI text-embeddings-3-small`, the embeddings' size is 1,536.

These embeddings are used in search algorithms. Azure AI Search uses an algorithm called **Hierarchical Navigable Small World** (**HNSW**), a flexible algorithm that's closely related to nearest neighbors for high-dimensional spaces, such as the number of dimensions of our embeddings. We'll use this algorithm later to search for items in our index and bring the ones that are more closely related. Let's add it to our embedding field:

```
    # Configure the vector search configuration
    vector_search = VectorSearch(
        algorithms=[
            HnswAlgorithmConfiguration(
                name="myHnsw",
                kind=VectorSearchAlgorithmKind.HNSW,
                parameters=HnswParameters(
                    m=10,
                    ef_construction=400,
                    ef_search=500,
                    metric=VectorSearchAlgorithmMetric.COSINE
                )
            )
        ],
```

```
        profiles=[
            VectorSearchProfile(
                name="myHnswProfile",
                algorithm_configuration_name="myHnsw",
            )
        ]
    )
```

In the preceding snippet, we used cosine similarity as the metric that determines the items in the index that are more closely related to what the user searched for. For now, we've used the default parameters of `m=10`, `ef_construction=400`, and `ef_search=500`. `ef` in the parameters stands for *exploration factor*.

The `m` parameter controls the density of the index – in the index, each record will have `m` neighbors. The `ef_construction` parameter increases the number of candidates being used to find neighbors for each record: the higher this parameter, the more thorough the search is going to be. The `ef_search` parameter controls the depth of the search during runtime – that is, when a search is executed, how many results are retrieved from the index for comparison.

Increasing `ef_construction` causes the index construction to take longer, whereas increasing `ef_search` causes runtime searches to take longer. In most cases, the numbers can be close to each other, but if you are planning to update the index frequently and don't want the construction time to take longer, you may increase `ef_search`. On the other hand, if your searches are already taking long enough at runtime and you want to improve their quality, you may increase `ef_construction` as it will make the results better and only increase the time it takes to build the index, but not the time it takes to execute a search.

Higher values for these parameters make the index better at finding records, but they also make it take longer to build and search through. The parameters we used here work well for this example, but when you are using your own dataset for your application, be sure to experiment with the parameters.

Finally, we simply call `create_or_update_index` with all the parameters we specified. This command is what will create the index:

```
    # Create the search index with the semantic settings
    index = SearchIndex(name=index_name, fields=fields,
                        vector_search=vector_search)
    result = index_client.create_or_update_index(index)
    print(f' {result.name} created')

if __name__ == '__main__':
    load_dotenv()
    main()
```

Now that we have an index, we can upload the records (each record is called a document) from our dataset into it.

Next, we'll learn how to create the index with C#.

Creating the index using C#

It's a lot simpler to create the index using C#. First, we must define the fields in a class, which I chose to call `SearchModel`:

```csharp
using Azure.Search.Documents.Indexes;
using Azure.Search.Documents.Indexes.Models;
public class SearchModel
{
    [SimpleField(IsKey = true, IsSortable = true, IsFilterable = true,
IsFacetable = true)]
    public string Id { get; set; }

    [SearchableField]
    public string AdditionalMetadata { get; set; }

    [SearchableField]
    public string Text { get; set; }

    [SearchableField]
    public string Description { get; set; }

    [SearchableField]
    public string ExternalSourceName { get; set; }

    [SimpleField(IsFilterable = true)]
    public bool IsReference { get; set; }

}
```

Here, we are using the same field names that we used for Python. Note that we didn't create an `Embedding` field like in Python. This will be created later, dynamically, when we load the documents.

Let's see how to create an index:

```csharp
using Azure;
using Azure.Search.Documents;
using Azure.Search.Documents.Indexes;
using Azure.Search.Documents.Indexes.Models;

var (apiKey, orgId, searchServiceName, searchServiceAdminKey,
```

```
searchIndexName) = Settings.LoadFromFile();

string indexName = searchIndexName;
AzureKeyCredential credential = new
AzureKeyCredential(searchServiceAdminKey);


SearchIndexClient indexClient = new SearchIndexClient(new
Uri(searchServiceName), credential);

indexClient.DeleteIndex(indexName);

var fields = new FieldBuilder().Build(typeof(SearchModel));

SearchIndex index = new SearchIndex(indexName)
{
    Fields = fields,
    // Add vector search configuration if needed
};

var result = indexClient.CreateOrUpdateIndex(index);
```

The code is straightforward: first, we create a list of fields in `SearchModel` using the `FieldBuilder` class; then, we create an `index` object with the `SearchIndex` class; and finally, we call `CreateOrUpdateIndex` to create the index in the cloud service.

Uploading documents to the index

While loading documents to the index is also straightforward, there are a couple of details that we need to pay attention to.

The first detail is the unique identifier of the document. In our case, that is the `Id` field. In an ideal case, the data that you want to load will have a unique and immutable identifier.

Luckily, that is the case for the ArXiV database: the `Id` field in the ArXiV database is unique and immutable and can always be used to search for articles online. For example, the article with an ID of `2309.12288` will always be the latest version of the *The Reversal Curse: LLMs trained on "A is B" fail to learn "B is A"* article [2], which talks about a quirk in LLMs: when asked who Tom Cruise's mother is, it will give the correct answer, Mary Lee Pfeiffer, 79% of the time. When asked who Mary Lee Pfeiffer's famous actor son is, it will give the correct answer, Tom Cruise, only 33% of the time.

The uniqueness and immutability of the `Id` field allow us to update the index with new information as needed. However, there's one caveat: in the index, the `Id` field can only contain numbers, letters, and underscores, so we will need to replace the dot with an underscore.

The second detail is that we need to load the embeddings. For Python, at the time of writing, this will require us to calculate the embeddings manually, as we did in *Chapter 6*. Different embedding models produce data vectors with different meanings, and usually different sizes, but even if the sizes are the same, the embeddings are incompatible unless explicitly stated.

Therefore, you can't create your embeddings with a model and later use another embedding model to do your searches. Also, this means that whoever is writing the code to perform searches needs to know the exact embedding model that was used to load the data in the index. In C#, we can use a connector called `Microsoft.SemanticKernel.Connectors.AzureAISearch`. That connector, while still in preview, will greatly simplify things. This should be available for Python soon but isn't at the time of writing.

Now that we know about these two details, let's write some code that will load the documents into the index.

Uploading documents with Python

We start by importing several packages:

```
from azure.core.credentials import AzureKeyCredential
from azure.search.documents import SearchClient
```

The first set of packages is for connecting to the Azure AI Search index. The packages are similar to the ones we used when creating the index, but note that we're using a different class, `SearchClient`, instead of `SearchIndexClient`.

Now, let's load the Semantic Kernel packages:

```
import asyncio
import semantic_kernel as sk
import semantic_kernel.connectors.ai.open_ai as sk_oai
```

These Semantic Kernel packages are going to be used to connect to the OpenAI service and generate the embeddings.

Finally, we're going to import some packages to help us control the flow of the program:

```
from tenacity import retry, wait_random_exponential, stop_after_
attempt
import pandas as pd
import os
from dotenv import load_dotenv
```

The `tenacity` library is helpful when you need to call functions that may fail as it provides you with functionality that allows you to automatically retry. The `pandas` library is used to load a CSV file. It's not strictly necessary; you can manipulate CSVs directly without it, but the `pandas` library makes it easier.

Next, let's define a helper function to generate embeddings:

```
@retry(wait=wait_random_exponential(min=1, max=5), stop=stop_after_
attempt(3))
async def generate_embeddings(kernel: sk.Kernel, text):
    e = await kernel.get_service("emb").generate_embeddings(text)
    return e[0]
```

This function assumes we have a kernel with a service named emb that can generate embeddings for a given text. We used the retry decorator to try to generate embeddings three times before giving up, waiting between 1 and 5 seconds between each try, increasing the interval as the number of tries increased.

Since the OpenAI service that we're going to use for generating embeddings is an online service and we have more than 30,000 articles to generate embeddings for, we are going to call it more than 30,000 times. With so many calls, it's not uncommon for some of them to occasionally fail due to network connectivity or the service being too busy. Therefore, adding the retry functionality can help so that you don't get an error on call number 29,000 that breaks your program.

> **Important – Using the OpenAI services is not free**
>
> To generate embeddings, we must call the OpenAI API. These calls require a paid subscription, and each call will incur a cost. The costs are usually small per request —version 3 of the embedding models costs $0.02 per million tokens at the time of writing this book, but costs can add up.
>
> OpenAI pricing details can be found at https://openai.com/pricing.
>
> Azure OpenAI pricing details can be found at https://azure.microsoft.com/en-us/pricing/details/cognitive-services/openai-service/.

The process we'll follow to create the index search client for loading documents is very similar to what we did when creating the index. The SearchClient class has one more parameter than SearchIndexClient, which we used to create the index: the index_name property that we created before:

```
async def main():
    kernel = sk.Kernel()
    api_key = os.getenv("OPENAI_API_KEY")
    embedding_gen = sk_oai.OpenAITextEmbedding(service_id="emb", ai_
model_id="text-embedding-3-large", api_key=api_key)
    kernel.add_service(embedding_gen)

    index_name = os.getenv("ARXIV_SEARCH_INDEX_NAME")
    service_name = os.getenv("ARXIV_SEARCH_SERVICE_NAME")
    service_endpoint = f"https://{service_name}.search.windows.net/"
    admin_key = os.getenv("ARXIV_SEARCH_ADMIN_KEY")
```

```
credential = AzureKeyCredential(admin_key)

# Create a search index
index_client = SearchClient(index_name=index_name,
    endpoint=service_endpoint, credential=credential)
```

Let's load the data:

```
df = pd.read_json('ai_arxiv_202101.json', lines=True)

count = 0
documents = []
for key, item in df.iterrows():
    id = str(item["Id"])
    id = id.replace(".", "_")
```

Here, we read the data file into a `pandas` DataFrame, and for each record, we create a dictionary called `document`. Note that we must replace periods with underscores in the `Id` field because Azure AI Search requires key fields to only contain numbers, letters, dashes, and underscores.

Now that we have the data in the dictionary, we are ready to upload it, which we will do in the following code:

```
        embeddings = await generate_embeddings(kernel,
  item["abstract"])
        # convert embeddings to a list of floats
        embeddings = [float(x) for x in embeddings]

        document = {
            "@search.action": "upload",
            "Id": id,
            "Text": item["title"],
            "Description": item["abstract"],
            "Embedding": embeddings
        }
        documents.append(document)
```

The fields in the document dictionary match the fields that we used when we created the index: `Id`, `Text`, `Description`, and `Embedding`. The value for the `Embedding` field is generated by calling the `generate_embeddings` function we created earlier.

Also, note the additional field, `@search.action`. This field contains instructions on what's going to happen with that item when it's submitted to the index. `upload` is a good default as it creates the record with that ID if it doesn't exist and updates its contents in the index if it does.

Lastly, once we've created the `document` dictionary item, we append it to the `documents` list.

Now, we are ready to upload it to the index:

```
N = 100
for i in range(0, len(documents), N):
    result = index_client.upload_documents(documents[i:i+N])
    print(f"Uploaded {len(documents[i:i+N])} records")

print(f"Final tally: inserted or updated {len(documents)}
records")
```

When uploading data to the index, there's a limit of 16 MB per operation. Therefore, we can only upload a few records at a time. In the preceding code, I limited the number of records uploaded to 100. However, any small enough number works since we are going to insert records into the index just once. The upload operation doesn't take very long, and it's better to upload a few records at a time and have a slightly longer upload duration than trying to upload many records at a time and risk getting an error.

The final step is to call the `main` function:

```
if __name__ == "__main__":
    load_dotenv()
    asyncio.run(main())
```

Note that before calling the `main` function, we called `load_dotenv` to get the values of the environment variables that contain the index name, the service name, the admin key, and the OpenAI key.

Running this program will cost approximately $1.50 as it will generate the embeddings. It will take about two and a half hours to run since we are generating dozens of thousands of embeddings. If you want to reduce the cost or time for your experiment, you can simply load just a fraction of the documents.

Once the program finishes running, you will see the following printed message:

```
Final tally: inserted or updated 35,808 records.
```

Now, we can use the index to find articles. Later, we will use it to answer questions about AI papers. But before we do that, let's learn how to upload the documents to the index using C#.

Uploading documents with C#

The `Microsoft.SemanticKernel.Connectors.AzureAISearch` package, which is in preview at the time of writing, makes it a lot easier to upload documents with C#. To use it, we must install it:

```
dotnet add package Microsoft.SemanticKernel.Connectors.AzureAISearch
--prerelease
```

Also, add the OpenAI connectors package:

```
dotnet add package Microsoft.SemanticKernel.Connectors.OpenAI
```

Now, we are going to use these packages to load the following documents into the index:

```
using Microsoft.SemanticKernel.Connectors.AzureAISearch;
using Microsoft.SemanticKernel.Memory;
using Microsoft.SemanticKernel;
using Microsoft.SemanticKernel.Connectors.OpenAI;
using System.Text.Json;
```

Since the package is in prerelease form, we need to add a few `pragma` directives to let C# know that we know that we are using prerelease functionality:

```
#pragma warning disable SKEXP0020
#pragma warning disable SKEXP0010
#pragma warning disable SKEXP0001
ISemanticTextMemory memoryWithCustomDb;
```

At this point, we can get our environment variables. I've modified the `Settings.cs` file to allow for the additional Azure AI Search variables to be stored and read from `config/settings.json`. For brevity, I won't put the file here, but you can check out this chapter's GitHub repository to see it:

```
var (apiKey, orgId, searchServiceName, searchServiceAdminKey,
searchIndexName) = Settings.LoadFromFile();
```

Next, we must create a `Memory` object with `MemoryBuilder`. We will use the `AzureAISearchMemoryStore` class to connect to Azure AI Search:

```
memoryWithCustomDb = new MemoryBuilder()
                .WithOpenAITextEmbeddingGeneration("text-embedding-3-
small", apiKey)
                    .WithMemoryStore(new
AzureAISearchMemoryStore(searchServiceName, searchServiceAdminKey))
                    .Build();
```

The next step is to read the data from the `ai_arxiv.json` file. Despite its extension, it's not a JSON file; it's a text file with one JSON object per line, so we will parse each line one at a time:

```
string data = File.ReadAllText("ai_arxiv.json");

int i = 0;
foreach (string line in data.Split('\n'))
{
    i++;
    var paper = JsonSerializer.Deserialize<Dictionary<string,
```

```
object>>(line);
    if (paper == null)
    {
        continue;
    }
    string title = paper["title"]?.ToString() ?? "No title available";
    string id = paper["id"]?.ToString() ?? "No ID available";
    string abstractText = paper["abstract"]?.ToString() ?? "No
abstract available";
    id = id.Replace(".", "_");
```

The next step is to use the `SaveInformationAsync` method of the `MemoryStore` object to upload the document into the index:

```
    await memoryWithCustomDb.SaveInformationAsync(collection:
searchIndexName,
        text: abstractText,
        id: id,
        description: title);

    if (i % 100 == 0)
    {
        Console.WriteLine($"Processed {i} documents at {DateTime.
Now}");
    }
}
```

Now that we've loaded the documents into the index, we can learn how to use the index to run a simple search. Later, we will use the results of the search and Semantic Kernel to assemble an answer.

Using the index to find academic articles

This subsection assumes that the index was loaded in the previous step. The index now contains titles, abstracts, and embeddings for thousands of academic papers about LLMs from ArXiV. Note that the papers in ArXiV are not necessarily peer-reviewed, which means that some articles may contain incorrect information. Regardless, ArXiV is generally a reputable data source for academic articles about AI, and many classic papers can be found there, including *Attention is All You Need* [3], the academic article that introduced GPT to the world. That article is not part of our dataset because our dataset starts in 2021, and that article is from 2017.

We're going to use this index to help us find papers for a given search string. This will ensure that the search is working and that we're comfortable with the results. In the next subsection, we're going to use GPT to combine the search results and summarize the findings. Let's see how to do this in Python and C#.

Searching for articles in Python

The first thing we must do is load the required libraries:

```
import asyncio
import logging
import semantic_kernel as sk
import semantic_kernel.connectors.ai.open_ai as sk_oai
from azure.core.credentials import AzureKeyCredential
from azure.search.documents import SearchClient
from azure.search.documents.models import VectorizedQuery
from dotenv import load_dotenv
from tenacity import retry, wait_random_exponential, stop_after_
attempt
import pandas as pd
import os


@retry(wait=wait_random_exponential(min=1, max=5), stop=stop_after_
attempt(3))
async def generate_embeddings(kernel: sk.Kernel, text):
    e = await kernel.get_service("emb").generate_embeddings(text)
    # convert e[0] to a vector of floats
    result = [float(x) for x in e[0]]
    return result
```

There are no new libraries, and we're using the same `generate_embeddings` function as we did before. The function that's used to generate embeddings when searching must be compatible with the function that was used to store embeddings in the vector database. If you use the same model, the function is going to be compatible.

In the following code, we're creating a kernel and loading the embedding model into it:

```
def create_kernel() -> sk.Kernel:
    kernel = sk.Kernel()
    api_key = os.getenv("OPENAI_API_KEY")
    embedding_gen = sk_oai.OpenAITextEmbedding(service_id="emb", ai_
model_id="text-embedding-3-small", api_key=api_key)
    kernel.add_service(embedding_gen)
    return kernel


async def main():
    kernel = create_kernel()

    ais_index_name = os.getenv("ARXIV_SEARCH_INDEX_NAME")
    ais_service_name = os.getenv("ARXIV_SEARCH_SERVICE_NAME")
```

```
    ais_service_endpoint = f"https://{ais_service_name}.search.
windows.net/"
    ais_admin_key = os.getenv("ARXIV_SEARCH_ADMIN_KEY")
    credential = AzureKeyCredential(ais_admin_key)
    search_client = SearchClient(ais_service_endpoint, ais_index_name,
credential=credential)
```

Besides loading the kernel with the embeddings model, we've also loaded all the environment variables with the configuration of our OpenAI connection and our Azure AI Search connection.

Now, we can execute the query:

```
    query_string = "<your query here>"
    emb = await generate_embeddings(kernel, query_string)
    vector_query = VectorizedQuery(vector=emb, k_nearest_neighbors=5,
exhaustive=True, fields="Embedding")

    results = search_client.search(
        search_text=None,
        vector_queries= [vector_query],
        select=["Id", "Text", "Description"]

    )
```

Executing this query consists of a few steps. First, we calculate the embeddings from our query string. This is done with the `generate_embeddings` function. Then, we create `VectorizedQuery` with the embeddings before executing the query using the `search` method of `SearchClient`. The `k_nearest_neighbors` parameter of our query determines how many results we want to bring back. In this case, I'm bringing back the first 5.

The results come in a dictionary with the columns in the index. We'll also retrieve an additional special column called `@search.score` that's created dynamically during the search and shows the cosine similarity of each result:

```
    pd_results = []
    for result in results:
        d = {
            "id": result['Id'],
            "title": result['Description'],
            "abstract": result['Text'],
            "score": f"{result['@search.score']:.2f}"
        }
        pd_results.append(d)
```

The values of the `@search.score` field may be used to sort results by order of similarity, and also to drop results that are below a cut-off point.

Let's print the results:

```
pd_results = pd.DataFrame(pd_results)

# print the title of each result
for index, row in pd_results.iterrows():
    print(row["title"])

if __name__ == "__main__":
    load_dotenv()
    asyncio.run(main())
```

In the preceding code, I'm loading the results into a `pandas` DataFrame before printing them as this makes it easier to sort and filter results when you have too many. This isn't required, though – you can simply use a dictionary. In this case, we're limiting our results to only five, so we could also print them directly from the `pd_results` dictionary list we created. For example, let's say we have the following query:

```
query_string = "models with long context windows lose information in
the middle"
```

We'll look at the results after we've learned how to implement the search in C#.

Searching for articles with C#

We can create our `memoryWithCustomDb` object in the same way we did to load the documents. Up to that point, the code is the same. However, instead of loading documents, we will now search for them. We can do that with the `SearchAsync` method of the `memoryWithCustomDb` object. All we need to do is pass the name of our index, which is stored in the `searchIndexName` variable from the configuration, the query we want to make, which we specified in `query_string`, and the number of articles we want to retrieve, which we specified in `limit`. We set `minRelevanceScore` to `0.0` so that we always retrieve the top five results. You can set it to a higher number if you only want to return results that exceed a minimum cosine similarity:

```
IAsyncEnumerable<MemoryQueryResult> memories = memoryWithCustomDb.
SearchAsync(searchIndexName, query_string, limit: 5,
minRelevanceScore: 0.0);

int i = 0;
await foreach (MemoryQueryResult item in memories)
{
    i++;
    Console.WriteLine($"{i}. {item.Metadata.Description}");
}
```

With the dedicated `memoryWithCustomDb` C# object, querying the memory is very simple, and we can get our results with a single `SearchAsync` call.

Let's check out the results.

Search results

For both Python and C#, the results we get are as follows:

```
1. Lost in the Middle: How Language Models Use Long Contexts
2. Parallel Context Windows for Large Language Models
3. Revisiting Parallel Context Windows: A Frustratingly Simple
Alternative and Chain-of-Thought Deterioration
4. "Paraphrasing The Original Text" Makes High Accuracy Long-Context
QA
5. Emotion Detection in Unfix-length-Context Conversation
```

Now that we have seen that the search works in C# and Python, we can use RAG to automatically generate a summary of several papers based on a search we'll make.

Using RAG to create a summary of several articles on a topic

We're going to use the search results from the previous step and add them to a prompt by using the usual semantic function prompt template. The prompt will instruct a model – in our case, GPT-4 – to summarize the papers our search returned.

Let's start with the semantic function, which we will call `summarize_abstracts`. Here's its metaprompt:

skprompt.txt

```
You are a professor of computer science writing a report about
artificial intelligence for a popular newspaper.
Keep the language simple and friendly.
Below, I'm going to give you a list of 5 research papers about
artificial intelligence. Each paper has a number and an abstract.
Summarize the combined findings of the paper. When using the
abstracts, refer to them by using their number inside [] brackets.
Your summary should be about 250 words.
Abstracts:
{{$input}}
```

The key part of the prompt is that when we ask for the summarization, I ask GPT to refer to the number of the abstract. For that to work, we will generate a list that has a number and an abstract, which is very similar to the results we generated in the *Using the index to find academic articles* section. The difference is that instead of having a number and the article title, we will have a number and the article abstract. Let's see the configuration file:

config.json

```
{
    "schema": 1,
    "type": "completion",
    "description": "Summarize abstracts of academic papers",
    "execution_settings": {
      "default": {
        "max_tokens": 4000,
        "temperature": 0.5
      }
    },
    "input_variables": [
      {
        "name": "input",
        "description": "A numbered list of abstracts to summarize.",
        "required": true
      }
    ]
}
```

Here, the most important thing you must do is make sure that you have enough tokens in the max_tokens field. You're going to be sending five abstracts, which might easily get to 200 tokens per abstract, so you need at least 1,000 tokens just for the abstracts and more for the instructions and the response.

Retrieving the data with Python and calling the semantic function

The first thing we need to do is add a generative model to our kernel. I've modified the create_ kernel function so that it adds a gpt-4-turbo model:

```
def create_kernel() -> sk.Kernel:
    kernel = sk.Kernel()
    api_key = os.getenv("OPENAI_API_KEY")
    embedding_gen = sk_oai.OpenAITextEmbedding(service_id="emb", ai_
model_id="text-embedding-3-small", api_key=api_key)
    gpt_gen = sk_oai.OpenAIChatCompletion(service_id="gpt-4-turbo",
ai_model_id="gpt-4-turbo-preview", api_key=api_key)
    kernel.add_service(gpt_gen)
    kernel.add_service(embedding_gen)
    return kernel
```

You can use any model you want, but I decided on gpt-4-turbo because it provides a good balance between cost and performance.

The next step is to create a function to summarize documents:

```
async def summarize_documents(kernel: sk.Kernel, df: pd.DataFrame) ->
str:
    doc_list = ""
    i = 0
    doc_list += "Here are the top 5 documents that are most similar to
your query:\n\n"
    for key, row in df.iterrows():
        i = i + 1
        id = row["Id"].replace("_", ".")
        doc_list += f"{i}. "
        doc_list += f"{row['Description']} - "
        doc_list += f"https://arxiv.org/abs/{id}\n"
```

The first part of the function creates a string that specifies a numbered list of documents and their URLs. Because of the way Azure AI Search stores IDs, remember that we had to convert dots into underscores. To generate the proper URL, we must convert it back.

The second part of the function generates a list of abstracts with the same numbers as the papers. When we write the prompt, we can ask the model to refer to the numbers, which, in turn, refer to the articles' titles and URLs:

```
    a = 0
    abstracts = ""
    for key, row in df.iterrows():
        a = a + 1
        abstracts += f"\n\n{a}. {row['Text']}\n"
```

The next step is to load the semantic function from its configuration directory:

```
    f = kernel.import_plugin_from_prompt_directory(".", "prompts")
    summary = await kernel.invoke(f["summarize_abstracts"],
input=abstracts)
```

The final step is to combine the list of papers and URLs with the generated summary and return it:

```
    response = f"{doc_list}\n\n{summary}"
    return response
```

Now that we know how to do the retrieval with Python, let's see how to do it with C#. We'll look at the results after.

Retrieving the data with C# and calling the semantic function

To retrieve the data, we'll start with the same code we used in the *Using the index to find academic articles* section until we fill the `memories` variable:

```
IAsyncEnumerable<MemoryQueryResult> memories = memoryWithCustomDb.
SearchAsync(searchIndexName, query_string, limit: 5,
minRelevanceScore: 0.0);
```

I've started the response by listing the documents that were retrieved, their numbers, and their URLs, all of which were built from the `Id` field. There's no need to use an AI model for this step:

```
string explanation = "Here are the top 5 documents that are most like
your query:\n";
int j = 0;
await foreach (MemoryQueryResult item in memories)
{
    j++;
    string id = item.Metadata.Id;
    id.Replace('_', '.');
    explanation += $"{j}. {item.Metadata.Description}\n";
    explanation += $"https://arxiv.org/abs/{id}\n";
}
explanation += "\n";
```

Then, instead of creating a string with all the titles, as we did in the *Using the index to find academic articles* section, we are going to create a string named `input` with the five abstracts, identified by a number in the `i` variable. This will be used as the input parameter for our semantic function:

```
string input = "";
int i = 0;
await foreach (MemoryQueryResult item in memories)
{
    i++;
    input += $"{i}. {item.Metadata.Text}";
}
```

Now, we can create a Semantic Kernel named `kernel`, add an AI service to it, and load the semantic function defined in the previous subsection, which I decided to call `rag`:

```
Kernel kernel = Kernel.CreateBuilder()
                        .AddOpenAIChatCompletion("gpt-4-turbo",
apiKey, orgId, serviceId: "gpt-4-turbo")
                        .Build();

var rag = kernel.ImportPluginFromPromptDirectory("prompts",
```

```
"SummarizeAbstract");

explanation += await kernel.InvokeAsync(rag["summarize_abstracts"],
new KernelArguments() {["input"] = input});

Console.WriteLine(explanation);
```

Let's run our programs and see the results we get when we use the same `query_string` value that we used for the test in the previous subsection.

RAG results

The query that we will use here is `"models with long context windows lose information in the middle"`.

The results aren't deterministic, but you should get something similar to the following:

```
Here are the top 5 documents that are most like your query:
1. Lost in the Middle: How Language Models Use Long Contexts -
https://arxiv.org/abs/2307.03172
2. Parallel Context Windows for Large Language Models - https://arxiv.
org/abs/2212.10947
3. Revisiting Parallel Context Windows: A Frustratingly Simple
Alternative and Chain-of-Thought Deterioration - https://arxiv.org/
abs/2305.15262
4. "Paraphrasing the Original Text" Makes High Accuracy Long-Context
QA - https://arxiv.org/abs/2312.11193
5. Emotion Detection in Unfix-length-Context Conversation - https://
arxiv.org/abs/2302.06029
In the rapidly evolving field of artificial intelligence, particularly
in the realm of language models, recent research has been shedding
light on both the capabilities and limitations of these advanced
systems. A critical challenge identified is the handling of long
text sequences by language models, which is essential for tasks such
as multi-document question answering and key-value retrieval [1].
Despite the advancements, it's observed that the performance of these
models often diminishes when they need to process relevant information
located in the middle of long contexts [1]. This indicates a need for
better strategies to enable models to effectively utilize long input
contexts.
To address these limitations, a novel method named Parallel Context
Windows (PCW) has been introduced, which allows off-the-shelf Large
Language Models (LLMs) to process long texts by dividing them into
smaller chunks. This method has shown substantial improvements in
handling diverse tasks requiring long text sequences without the
need for further training [2]. However, further analysis reveals that
PCW may not consistently enhance the models' understanding of long
contexts in more complex reasoning tasks, suggesting that the method's
design might not guarantee significant improvements in practical
applications [3].
```

> Another approach to enhancing long-context capabilities involves focusing on the quality of training data. It has been found that "effective" data, which can be achieved through techniques such as original text paraphrasing, is crucial for training models to handle long texts, leading to state-of-the-art performance in multi-document retrieval and question answering tasks [4].
>
> Additionally, research into variable-length context windows for predicting emotions in conversations introduces new modules to better capture conversational dynamics. This approach significantly outperforms existing models by more accurately determining the relevant context for emotion prediction [5].
>
> Collectively, these studies highlight the importance of innovative methods and quality training data in overcoming the challenges of processing long texts. They also underscore the need for continued exploration to enhance the practical applicability of language models in real-world scenarios.

As you can see, the summary is comprehensive, captures the main idea of each paper, and shows how the papers relate to one another.

In this section, we learned how to use an external database to help an LLM work with a lot more information than the model can handle with its context window.

One advantage of this method is that the generative model primarily uses the search data that you supplied to it to generate a response. If you only supply it with real, well-curated data, you will substantially lower the chance that it will *hallucinate* – that is, generate information that doesn't exist. If you did not use RAG, there's a possibility that the generative model will make up non-existent papers and references just to try to answer the questions and generate the summaries we're asking for.

Blocking hallucinations completely is theoretically impossible, but using RAG can make the chance of hallucinating so low that, in practice, your users may never see fake data being generated by your model. This is the reason RAG models are used extensively in production applications.

Summary

In this chapter, we greatly expanded the data that's available to our AI models by using the RAG methodology. Besides allowing AI models to use large amounts of data when building prompts, the RAG methodology also improves the accuracy of the model: since the prompt contains a lot of the data that's required to generate the answer, models tend to hallucinate less.

RAG also allows AI to provide references to the material it used to generate a response. Many real-world use cases require models to manipulate large quantities of data, require references to be provided, and are sensitive to hallucinations. RAG can help overcome these issues easily.

In the next chapter, we will change gears and learn how to integrate a Semantic Kernel application with ChatGPT, making it available to hundreds of millions of users. In our example, we will use the application we built in *Chapter 5* for home automation, but you can use the same techniques to do that with your own applications.

References

[1] N. F. Liu et al., "Lost in the Middle: How Language Models Use Long Contexts." arXiv, Nov. 20, 2023. doi: 10.48550/arXiv.2307.03172.

[2] L. Berglund et al., "The Reversal Curse: LLMs trained on 'A is B' fail to learn 'B is A.'"   arXiv, Sep. 22, 2023. doi: 10.48550/arXiv.2309.12288.

[3] A. Vaswani et al., "Attention Is All You Need," Jun. 2017.

8

Real-World Use Case – Making Your Application Available on ChatGPT

In earlier chapters, we learned quite a lot. We learned how to create and optimize prompts, how to create semantic and native functions and put them in Semantic Kernel, and how to use a planner to automatically decide which functions of the kernel to use to solve a user problem.

In the previous two chapters, we learned how to augment our kernel with **memories**, including memories built from external data, which allows us to build more personalized applications and use data that is recent and that we have control over to generate answers, instead of using only the data that was used to train the LLM, which is frequently not public.

In this final chapter, we will change gears. Instead of creating new functionality, we will learn how to make the functionality we have already created available for many more users. We will use the home automation application that we wrote in *Chapter 5* and make it available through the OpenAI custom **GPT Store**, making it available to the hundreds of millions of users that already use ChatGPT, and use ChatGPT as the UI of our application.

Besides the obvious benefits of quickly being able to make an application available to hundreds of thousands of users, another benefit is that you don't even need to build a **user interface** (**UI**) for your application. You can build the main functionality and use ChatGPT as the UI. Of course, this has limitations. The AI is text based and you have little control over it, but on the other hand, you can test and deploy your application a lot faster, and build a dedicated UI later.

In this chapter, we'll be covering the following topics:

- Creating a custom GPT in the OpenAI store
- Creating a web API wrapper for an application developed with Semantic Kernel
- Connecting the custom GPT with the OpenAI store through the web API wrapper

By the end of the chapter, you will have an application that is available to all ChatGPT users.

Technical requirements

To complete this chapter, you will need to have a recent, supported version of your preferred Python or C# development environment:

- For Python, the minimum supported version is Python 3.10, and the recommended version is Python 3.11

- For C#, the minimum supported version is .NET 8

In this chapter, we will call OpenAI services. Given the amount that companies spend on training these LLMs, it's no surprise that using these services is not free. You will need an **OpenAI API** key, obtained either directly through **OpenAI** or **Microsoft**, via the **Azure OpenAI** service.

If you are using .NET, the code for this chapter is at `https://github.com/PacktPublishing/Building-AI-Applications-with-Microsoft-Semantic-Kernel/tree/main/dotnet/ch8`.

If you are using Python, the code for this chapter is at `https://github.com/PacktPublishing/Building-AI-Applications-with-Microsoft-Semantic-Kernel/tree/main/python/ch8`.

To create your custom GPT, you will need an account with OpenAI.

You can install the required packages by going to the GitHub repository and using the following: `pip install -r requirements.txt`.

Custom GPT agents

On November 6, 2023, OpenAI introduced functionality that allows users to create custom, personalized versions of ChatGPT. These custom GPTs created by users can be shared with other users through OpenAI's GPT Store. This allows users without programming experience to add functionality to ChatGPT by simply writing instructions in natural language, and it also allows users with programming experience to connect ChatGPT to their applications, making such applications available to hundreds of millions of users.

Initially, these were called "custom GPTs," but now they are simply called GPTs. That may be confusing since the transformer technology used in most AI models is called **generative pre-trained transformer** (**GPT**), and the OpenAI implementation of these models is also called GPT with a version, such as GPT-3.5 or GPT 4.

In this section, when we use the name "GPT," unless otherwise noted, it means the custom GPTs that you can create inside of ChatGPT.

These GPTs can use custom prompts, such as the ones we use in semantic functions, and additional data, such as what we use in RAG models. You can add custom prompts and documents to your custom GPT by using a web interface, which we will show in the next subsection.

In addition, you can also allow your GPT to call external functions through a web API. Many companies created these interfaces and connected them to custom GPTs, such as Wolfram (the creators of the scientific software Mathematica), design companies such as Canva and Adobe, and many others.

In this section, as we did *Chapter 5*, imagine you work for a home automation company that has a product that allows someone to control their home through a device in their house, and now you want to allow them to do it with ChatGPT. We created the native function for this in *Chapter 5*, and in this chapter, we are going to use Microsoft Semantic Kernel tools to make that functionality available for ChatGPT users.

Before we start that more complex example, let's first create a simpler custom GPT just to familiarize ourselves with the process.

Creating a custom GPT

To create a GPT, you can navigate to `https://chat.openai.com/gpts/` and click the **Create** button on the top-right corner, or navigate directly to `https://chat.openai.com/gpts/ editor`. This will open a web interface that allows you to create a GPT. As you'd expect, you can create the GPT simply by chatting with ChatGPT. You can add custom instructions, specify the tone of the answers, and much more.

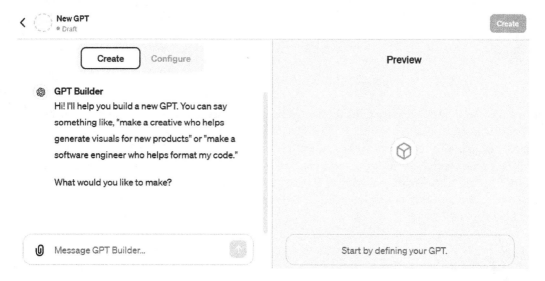

Figure 8.1 – Creating a GPT using the OpenAI editor

The **Configure** tab is where you will give your GPT a name and description, and where you can add custom actions that connect your GPT with external APIs:

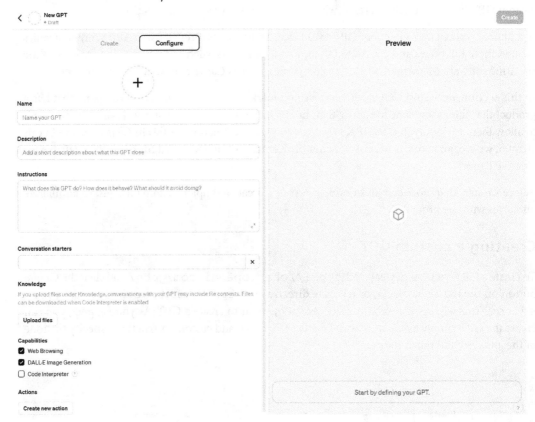

Figure 8.2 – The OpenAI UI to configure your GPT

You can use the **Create** interface to fill the configuration interface. For an initial example, I will build a GPT that answers questions about the Sherlock Holmes books. The first thing I told ChatGPT was I want to make a GPT that answers questions about Sherlock Holmes and it answered back with How about naming it Detective Guide? Does that sound good to you?. I answered back with Yes, and the configuration was updated, adding Detective Guide as the name. Without asking, ChatGPT also automatically generated an appropriate profile picture for my GPT.

I did a few more configuration steps:

```
I want it to be only about the canonical works by Arthur Conan Doyle,
but I want it to be helpful for high school students that are trying
to understand it better - the influence of the works in modern media
and the context in which the work was written.
```

I also uploaded the text files with the contents of the books, which you can find online and also in `/data/sherlock_holmes/`, and added this additional configuration:

```
When providing answers about the books, I want the gpt to base the
answers on the files I have uploaded and quote passages found in them.

Do not answer any questions that are not related to the Sherlock
Holmes' books. If a user asks questions about other topics, answer
that this is not your area of expertise.
```

After every configuration step, ChatGPT asks me if there's anything more I want to add. At this point, I said no, which enabled the **Create** button on the top-right corner again. Once I pressed it, I got the option to share my GPT:

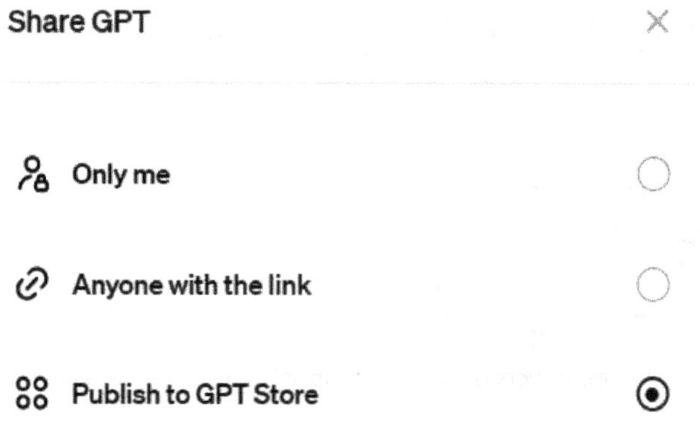

Share GPT   ✕

∻  Only me   ○

⟲   Anyone with the link   ○

■■   Publish to GPT Store   ◉

Figure 8.3 – Sharing your GPT in the GPT Store

I chose **Publish to GPT Store**, which will make it available to all ChatGPT users. The completed version, created only with the preceding minimal instructions, can be accessed at `https://chat.openai.com/g/g-yJvyO2vHq-detective-guide`.

This is what it looks like:

Figure 8.4 – The initial version of the Sherlock Holmes GPT, Detective Guide

One thing to notice is what little configuration I had to perform. I uploaded a few text files with the text of the books and wrote a few hundred words of configuration in natural language.

To test it, I asked What's Holmes attic theory?, and its answer, shown in the following, answers the question and correctly identifies the book it comes from:

```
Holmes' attic theory, a metaphorical description of how one should
manage their mental storage, emphasizes the importance of selective
knowledge acquisition. In "A Study in Scarlet," Holmes explains that
he considers the brain to be like a little empty attic, which one
must carefully furnish with only the tools that may aid one's work.
He argues against cluttering the mind with irrelevant information,
suggesting that such clutter can push out or obscure useful knowledge.
According to Holmes, it is crucial to maintain a well-organized and
selective collection of knowledge to ensure that one's mental capacity
is efficiently utilized for one's purposes.
```

> This theory sheds light on Holmes' approach to his detective
> work, where he values relevant, useful knowledge over a broad but
> shallow pool of information. It also underlines his practical view
> of intelligence, emphasizing application over accumulation. This
> perspective is central to understanding Holmes' character and his
> methods of deduction, which rely on connecting seemingly disparate
> facts stored neatly in his mental "attic" to solve complex cases.

In addition, because I asked the GPT to not answer questions outside of its knowledge domain, it tries to keep things on topic, as the following dialogue exemplifies:

Does the earth revolve around the sun?

Yes, the Earth revolves around the Sun. This is a well-established scientific fact that forms the basis of our understanding of the solar system. This understanding was significantly advanced by the work of astronomers like Nicolaus Copernicus in the 16th century, who proposed the heliocentric model of the solar system, and later by Johannes Kepler and Galileo Galilei.

In the context of Sherlock Holmes, there is an interesting conversation related to this topic in "A Study in Scarlet." In the book, Sherlock Holmes explains to Dr. Watson that he does not care whether the Earth revolves around the Sun or not because such knowledge does not affect his work as a detective:

> "What the deuce is it to me?" he interrupted impatiently; "you say that we go round the sun. If we went round the moon it would not make a pennyworth of difference to me or to my work".

This illustrates Holmes's practical approach to knowledge, focusing only on information directly useful to his detective work.

Figure 8.5 – Asking the GPT about astronomy and Sherlock Holmes' astronomy knowledge

> **If it's so easy to create a custom GPT, why write any code?**
>
> When GPT models started to become mainstream, several entrepreneurs created applications that were nothing more than GPT-3.5 with a few additional instructions. These apps were released on the web, like what we just did with the Detective Guide.
>
> If the functionality of your application can be replicated by adding a few additional instructions to ChatGPT, creating a custom GPT may be a good option for you. If you try to release it as a separate app and charge users, it's possible for a competitor to replicate your success by simply creating a custom GPT and offering it to all ChatGPT users. Monetization for these GPTs is still unclear, but apparently, it will work in the same way as Spotify or Kindle Unlimited: GPTs that get enough users receive a fraction of the money paid by subscribers.
>
> There are a few cases in which these custom GPTs in ChatGPT don't work at all. For example, you can't use it to add AI functionality to your existing application. You also don't have a lot of control: custom GPTs don't allow you to fine-tune your prompts, process the inputs and outputs of functions, monitor usage, change the sensitivity of search when retrieving documents, and so on, so you have a lot less control over the user experience.

Now that we have created a simple version of a custom GPT, let's create one that supports actions. We are going to use the home automation native plugin that we created in *Chapter 5* and pretend that we have a company that allows users to automate some functions in their house, such as turning the lights on and off and turning on the TV. We're going to create a custom GPT that allows users to use ChatGPT to control the functionality of their house.

Creating a custom GPT that supports actions

Like what we have done in the previous subsection, we start by navigating to `https://chat.openai.com/gpts/` and clicking the **Create** button in the top-right corner, or navigating directly to `https://chat.openai.com/gpts/editor`.

Instead of using the chat interface to create this custom GPT, I went directly to the **Configure** tab and added the following instructions:

```
You are designed to assist users with specific home automation tasks
through a web API, focusing on operating the garage door and turning
lights on and off. If users inquire about functionalities outside
these areas, you inform them that those actions are not supported.
When you need more information to provide accurate assistance, you
will ask for clarification directly. You strive to personalize your
responses to fit users' specific contexts and ensure clarity in your
explanations. Your interactions are professional and straightforward,
aiming for clarity and efficiency in communication.

You only work with the SKB custom API and no other vendors. If a user
asks about information about other vendors of home automation, say
that you can't help them.
```

I also named my GPT `SKB Home Automation Example`, where **SKB** stands for **Semantic Kernel Book**. You can leave most other options as their defaults, although you likely want to uncheck **Web Browsing** and **DALL-E Image Generation** under **Capabilities**, as these won't be used.

Once you are done, you can click **Create** on the top-right side of the screen, and this will create a custom GPT, which you can choose to keep to yourself, share with a link, or share in the GPT Store.

Whatever you choose, now you can send requests to the GPT. Note that we have not connected our native plugin to the GPT yet, so if you ask the GPT to perform something, such as `open the garage door`, it will immediately hallucinate, as shown here:

```
To open your garage door using the SKB Home Automation API, you would
typically send a command through a web request. The specific details
can depend on how your system is set up, but here's a general example
of how you might do it:

1. Ensure Your Device is Connected: Make sure your smartphone or
computer is connected to the internet and can communicate with your
home network where your garage door is connected.

2. Use the Right Endpoint: The SKB API will have a specific endpoint
for operating the garage door. This is often something like /api/
garage_door/open, but you should refer to the SKB API documentation
for the exact endpoint.

3. Send the Request: You can use various tools or a piece of code to
send the request. If you're comfortable with command line tools, you
might use curl in a terminal or command prompt like this:

   curl -X POST "http://yourhomeip:port/api/garage_door/open" -H
"Authorization: Bearer YOUR_API_TOKEN"

   Replace yourhomeip:port with the IP address and port number of your
SKB Home Automation system, and YOUR_API_TOKEN with your personal API
token for authentication.

4. Check the Response: After sending the request, you should receive
a response from the API. A successful response usually includes
a message indicating that the command was received and the action
(opening the garage door) is in progress or completed. If there's
an error, the response will typically contain details to help
troubleshoot the issue.

If you haven't set up your API token or aren't sure about the endpoint
or how to send the request, let me know what step you're on, and I can
provide more detailed instructions!
```

ChatGPT makes up several things in this answer: an API, a `curl` command, an authentication method, whether you can send commands to an IP address, and so on. None of those things are true; it's just trying its best to provide an answer. This problem will be solved when we connect ChatGPT with our real API, which we are going to do in the next section.

> **A note about security**
>
> When you share your GPT and your API with hundreds of millions of users, there's a real possibility that some of them will use it in ways that you have not thought about. We will not cover security in detail in this book, but this doesn't mean that you should not think about it.
>
> In the next section, we will connect an API to ChatGPT without any authentication, but for a production application, you should add authentication to the connection between GPT and your API. Most importantly, you should add monitoring to your API, so you can see if usage patterns are changing.
>
> Even the most basic monitoring that just counts how many calls you have had per minute is likely sufficient to prevent the worst forms of abuse. Once you have monitoring, you can also add rate limiting, to prevent malicious users from overwhelming your API with repeated calls.

Creating a web API wrapper for the native function

First, let's define our native function. This is the same function I used in *Chapter 5*, but I used only `OperateLight` and `OperateGarageDoor` for brevity:

```
import semantic_kernel as sk
from typing import Annotated
from semantic_kernel.functions.kernel_function_decorator import
kernel_function

class HomeAutomation:
    def __init__(self):
        pass

    @kernel_function(
        description="Turns the lights of the living room, kitchen,
bedroom or garage on or off.",
        name="OperateLight",
    )
    def OperateLight(self,
    location: Annotated[str, "The location where the lights are to be
turned on or off. Must be either 'living room', 'kitchen', 'bedroom'
or 'garage'"],
    action: Annotated[str, "Whether to turn the lights on or off"]) ->
Annotated[str,  "The output is a string describing whether the lights
were turned on or off" ]:
        if location in ["kitchen", "living room", "bedroom",
"garage"]:
            result = f"Changed status of the {location} lights to
{action}."

            return result
```

```
        else:
            error = f"Invalid location {location} specified."
            return error

    @kernel_function(
        description="Opens or closes the garage door.",
        name="OperateGarageDoor",
    )
    def OperateGarageDoor(self,
            action: Annotated[str, "Whether to open or close the
garage door"]) -> Annotated[str, "The output is a string describing
whether the garage door was opened or closed" ]:
        result = f"Changed the status of the garage door to {action}."
        return result
```

Now, we need to build a web API to make that function callable from the web by ChatGPT.

Creating a web API wrapper in Python

In Python, we are going to use the Flask library. In Flask, we're going to create two routes: `operate_light` and `operate_garage_door`. First, we create an app:

```
from flask import Flask, render_template, request, jsonify
from dotenv import load_dotenv
from semantic_kernel.connectors.ai.open_ai import OpenAIChatCompletion
import semantic_kernel as sk
from HomeAutomation import HomeAutomation

app = Flask(__name__)
app.secret_key = b'skb_2024'
```

Creating the app is simple, requiring just the calling of the `Flask` constructor and setting a `secret_key` property that can be used to sign cookies coming from your app. This app will not have cookies, so the secret key can be anything, including a random string.

Now, we will define the routes for our API:

```
@app.route('/operate_light', methods=['POST'])
async def operate_light():
    kernel = sk.Kernel()
    api_key, org_id = sk.openai_settings_from_dot_env()
    gpt4 = OpenAIChatCompletion("gpt-4-turbo-preview", api_key, org_
id)
    kernel.add_service(gpt4)
    kernel.import_plugin_from_object(HomeAutomation(),
```

```
"HomeAutomation")

    data = request.get_json()
    location = data['location']
    action = data['action']

    result = str(kernel.invoke(kernel.plugins["HomeAutomation"]
["OperateLight"], location=location, action=action))
    return jsonify({'result': result})

@app.route('/operate_garage_door', methods=['POST'])
async def operate_garage_door():
    kernel = sk.Kernel()
    api_key, org_id = sk.openai_settings_from_dot_env()
    gpt4 = OpenAIChatCompletion("gpt-4-turbo-preview", api_key, org_
id)
    kernel.add_service(gpt4)
    kernel.import_plugin_from_object(HomeAutomation(),
"HomeAutomation")

    data = request.get_json()
    action = data['action']
    result = str(kernel.invoke(kernel.plugins["HomeAutomation"]
["OperateGarageDoor"], action=action))
    return jsonify({'result': result})
```

The structure of each route is the same: we create a kernel, add a GPT service to it, import the HomeAutomation plugin, and invoke the appropriate function, returning its answer.

You can add these two lines of code to the application to allow for local testing:

```
if __name__ == '__main__':
    app.run()
```

To test the application locally, go to a command line and type the following:

```
flask run
```

This will create a local web server:

```
* Debug mode: off
WARNING: This is a development server. Do not use it in a production
deployment. Use a production WSGI server instead.
  * Running on http://127.0.0.1:5000
```

Now, you can send commands to the local web server using `curl` if you are using bash, or `Invoke-RestMethod` if you are using PowerShell. Here, we are invoking the `operate_light` route with `"action": "on"` and `"location": "bedroom"`:

```
Invoke-RestMethod -Uri http://localhost:5000/operate_light -Method
Post -ContentType "application/json" -Body '{"action": "on",
"location": "bedroom"}'
```

The result, as expected, is that the application responds successfully:

```
Result
------
Changed status of the bedroom lights to on.
```

Now that we verified that the web application is working, we can deploy it on the web.

Creating a web API wrapper in C#

.NET makes it easy to create a boilerplate web API application. You can use the following command and it will create a web API under the `SkHomeAutomation` directory:

```
dotnet new webapi --use-controllers -o SkHomeAutomation
```

Don't forget to install the `Microsoft.SemanticKernel` package, too:

```
dotnet add package Microsoft.SemanticKernel
```

The `dotnet new webapi` command helpfully generates the code for a weather forecasting web application that provides a web API. One of the files it generates is a module called `WeatherForecast.cs`. You can delete this file, as we will replace it with our own functionality. To do so, copy the `HomeAutomation.cs` file from *Chapter 5* to the root of this project. To make our life easier, add the following line to the beginning of the file, which will allow you to reference the `HomeAutomation` object more easily:

```
namespace SkHomeAutomation;
```

The last thing you need to do is to go into the `Controllers` directory. It will contain a `WeatherForecastController.cs` file. You can delete this file and replace it with the `HomeAutomationController.cs` file, which is here:

```
using Microsoft.AspNetCore.Mvc;
namespace SkHomeAutomation.Controllers;
using Microsoft.Extensions.Logging;

public class LightOperationData
{
    public string? location { get; set; }
    public string? action { get; set; }
}

public class GarageOperationData
{
    public string? action { get; set; }
}

[ApiController]
[Route("[controller]")]
public class HomeAutomationController : ControllerBase
{

    private readonly ILogger<HomeAutomationController>? _logger;
    private HomeAutomation ha;

    public HomeAutomationController(ILogger<HomeAutomationController>
logger)
    {
        _logger = logger;
        ha = new HomeAutomation();
    }

    [HttpPost("operate_light")]
    public IActionResult OperateLight([FromBody] LightOperationData
data)
    {
        if (data.location == null || data.action == null)
```

```
        {
            return BadRequest("Location and action must be provided");
        }
        return Ok( ha.OperateLight(data.action, data.location) );
    }


    [HttpPost("operate_garage_door")]
    public IActionResult OperateGarageDoor([FromBody]
GarageOperationData data)
    {
        if (data.action == null)
        {
            return BadRequest("Action must be provided");
        }
        return Ok( ha.OperateGarageDoor(data.action) );
    }


}
```

HomeAutomationController exposes the operate_light and operate_garage_door web API paths, and when those are called, it routes the request to the corresponding method of the HomeAutomation class that we created in *Chapter 5*, essentially exposing our Semantic Kernel application to the web, once it's deployed.

The next step, whether you created the application in C# or Python, is to deploy the application.

Deploying your application to an Azure Web App

To deploy your application on the web, you need to have an Azure account. Go to the Azure portal at https://portal.azure.com, and from the home page, click **Create a Resource** and then **Create a Web App**. As you will see here, we can use the free tier for our test, but if you plan to deploy something like this for a real application, you should choose a different plan.

In *Figure 8.6*, I show how I created mine: I created a new resource group called skb-rg, named my application skb-home-automation, which gives it the skb-home-automation. azurewebsites.net URL, and chose Python 3.11 (Python) or .NET 8 LTS (C#) for its runtime stack.

Under **Pricing plans**, I created a new **Linux Plan** called skb-sp, and chose the **Free F1** pricing plan. Once these configurations are done, click **Review + create** and your web application will be deployed in a few minutes:

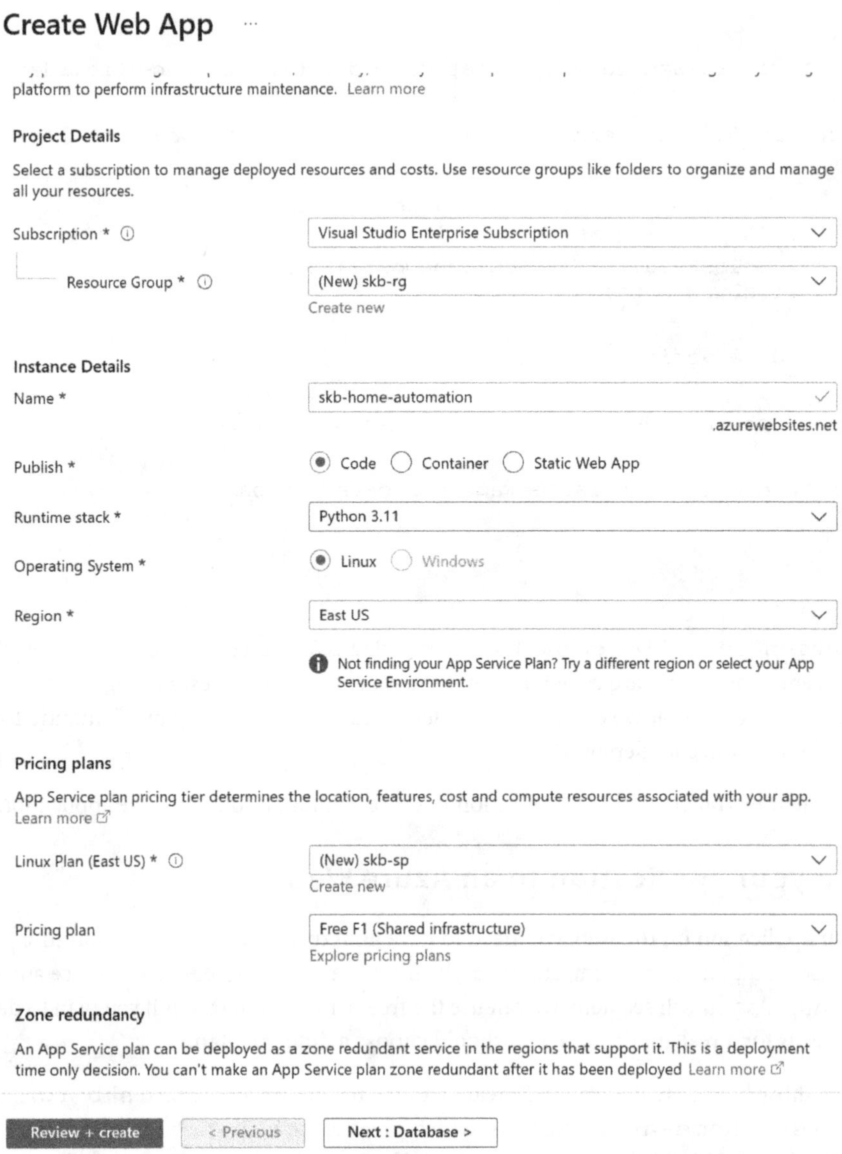

Create Web App   ···

platform to perform infrastructure maintenance. Learn more

Project Details

Select a subscription to manage deployed resources and costs. Use resource groups like folders to organize and manage all your resources.

Subscription * ⓘ	Visual Studio Enterprise Subscription ∨
└─ Resource Group * ⓘ	(New) skb-rg ∨
	Create new

Instance Details

Name *	skb-home-automation ✓
	.azurewebsites.net
Publish *	◉ Code ◯ Container ◯ Static Web App
Runtime stack *	Python 3.11 ∨
Operating System *	◉ Linux ◯ Windows
Region *	East US ∨

ⓘ Not finding your App Service Plan? Try a different region or select your App Service Environment.

Pricing plans

App Service plan pricing tier determines the location, features, cost and compute resources associated with your app. Learn more ☐

Linux Plan (East US) * ⓘ	(New) skb-sp ∨
	Create new
Pricing plan	Free F1 (Shared infrastructure) ∨
	Explore pricing plans

Zone redundancy

An App Service plan can be deployed as a zone redundant service in the regions that support it. This is a deployment time only decision. You can't make an App Service plan zone redundant after it has been deployed Learn more ☐

[Review + create]   [< Previous]   [Next : Database >]

Figure 8.6 – Creating a free web app to host our API

The simplest way to deploy your API to the web application is through GitHub. To do so, we need to create a new, clean GitHub repository for this web API and copy the contents of `https://github.com/PacktPublishing/Microsoft-Semantic-Kernel/tree/main/python/ch8` to it. This needs to be a separate repository because you need to deploy the full repository to the web application. For example, you can put your copy at an address such as `https://github.com/<your-github-username>/skb-home-automation`.

In your web application, go to **Deployment Center**, and select **GitHub** as the source. In **Organization**, select your username. Choose the repository.

This will create and deploy the web API under your own account.

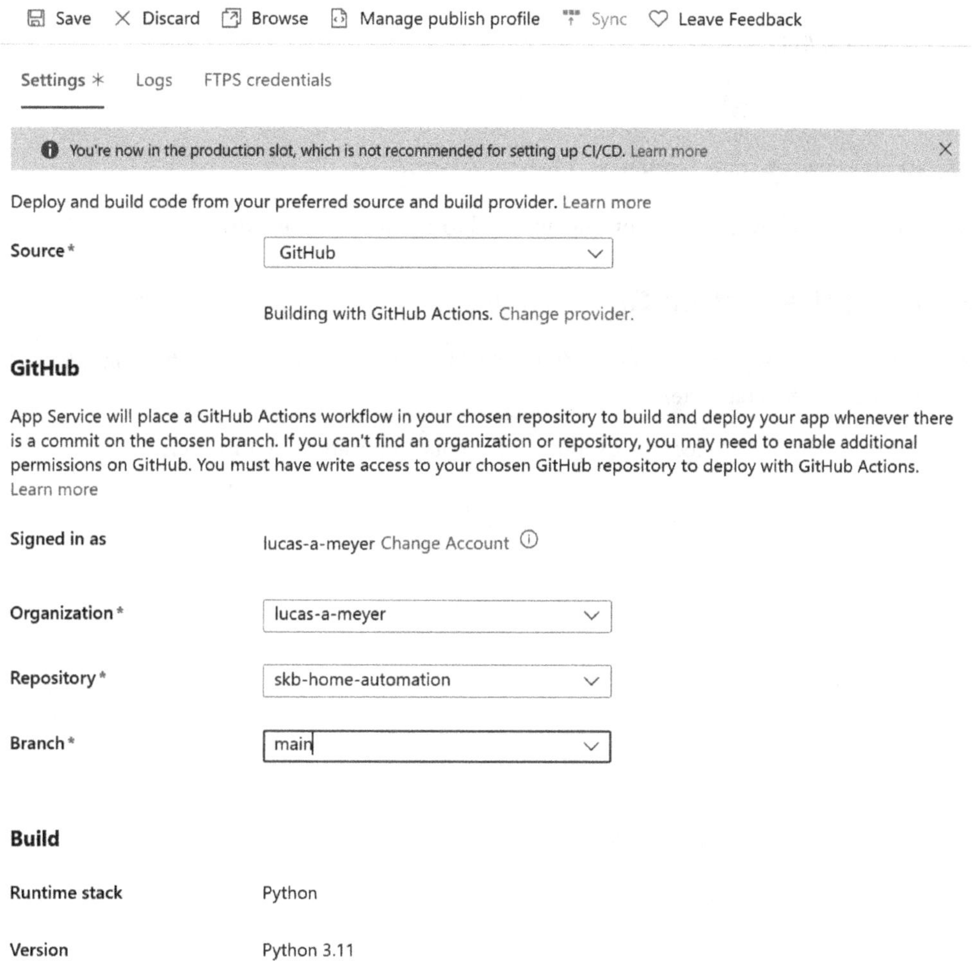

Figure 8.7 – Deploying the web API using GitHub

Once the web API is deployed, you can test it using `curl` or `Invoke-RestApi`. The only change is that instead of using localhost as the endpoint, you need to change it to the endpoint you deployed to. In my case, I chose `skb-home-automation.azurewebsites.net` (your case will be different). Please note that my version of the API will not be available for you; you must deploy your own.

Therefore, we can submit the following:

```
Invoke-RestMethod -Uri https://skb-home-automation.azurewebsites.
net/operate_light -Method Post -ContentType "application/json" -Body
'{"action": "on", "location": "bedroom"}'
```

The result will be as follows:

```
Result
------
Changed status of the bedroom lights to on.
```

Now that we have a web API that is working, we need to connect the API with ChatGPT.

Connecting the custom GPT with your custom GPT action

To connect our web API with our custom GPT, we need to give it an OpenAPI specification. ChatGPT makes it very easy to generate one.

First, go to our custom GPT, select the dropdown from its name, and select **Edit GPT**:

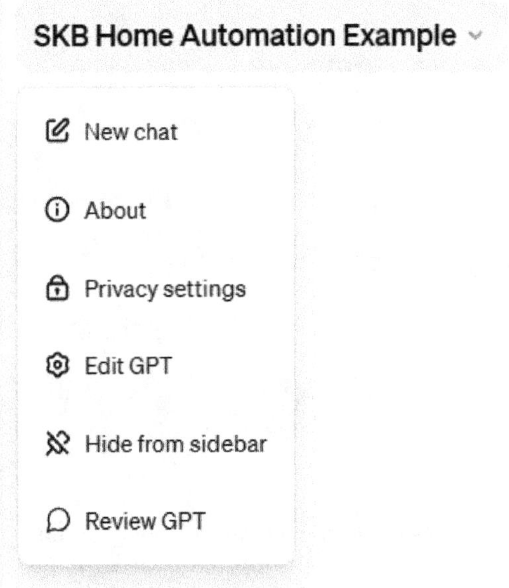

SKB Home Automation Example ∨

- ✏️ New chat
- ⓘ About
- 🔒 Privacy settings
- ⚙️ Edit GPT
- ✂️ Hide from sidebar
- 💬 Review GPT

Figure 8.8 – Editing our GPT

On the bottom of the **Configuration** tab, click on **Create new action**, under **Actions**. This will bring up the **Add actions** UI:

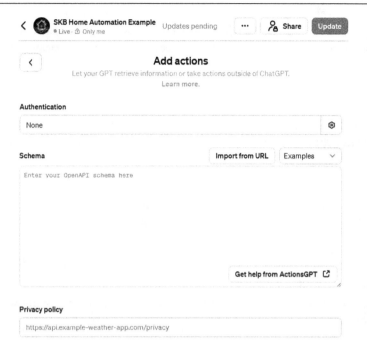

Figure 8.9 – Adding actions to our GPT

To add actions, you need to specify a schema using a language called **OpenAPI**. ChatGPT makes this extremely easy: clicking on **Get help from ActionGPT** will open a chat dialog with another custom GPT that can help you create OpenAPI specifications:

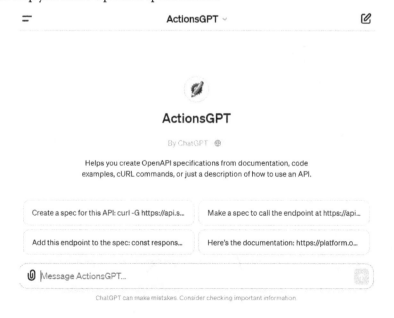

Figure 8.10 – Using ActionsGPT

In ActionsGPT, all you need to do is paste the code of our web API, and it will automatically generate the OpenAPI specification. What was automatically generated is in the following, but we'll need to make a couple of changes:

```yaml
openapi: 3.0.0
info:
  title: Home Automation API
  description: This API allows controlling lights and garage doors in
a home automation system.
  version: 1.0.0
servers:
  - url: http://yourserver.com
    description: Main server
paths:
  /operate_light:
    post:
      operationId: operateLight
      summary: Controls a light in the home automation system.
      requestBody:
        required: true
        content:
          application/json:
            schema:
              type: object
              required:
                - location
                - action
              properties:
                location:
                  type: string
                  description: The location of the light to be
controlled.
                action:
                  type: string
                  description: The action to be performed on the
light.
                  enum:
                    - turnOn
                    - turnOff
                    - toggle
      responses:
        '200':
          description: Operation result
          content:
```

```
              application/json:
                schema:
                  type: object
                  properties:
                    result:
                      type: string
                      description: The result of the light operation.


  /operate_garage_door:
    post:
      operationId: operateGarageDoor
      summary: Controls the garage door in the home automation system.
      requestBody:
        required: true
        content:
          application/json:
            schema:
              type: object
              required:
                - action
              properties:
                action:
                  type: string
                  description: The action to be performed on the
garage door.
                  enum:
                    - open
                    - close
                    - stop
      responses:
        '200':
          description: Operation result
          content:
            application/json:
              schema:
                type: object
                properties:
                  result:
                    type: string
                    description: The result of the garage door
operation.
```

Notably, it doesn't know the name of my server or the restrictions of the places where lights are installed. It also tries to guess the commands. Therefore, we must add the correct restrictions to the specification. Another detail to note is that I have the x-openai-isConsequential: false parameter for all endpoints. When that parameter is true or blank, ChatGPT will ask for confirmation for each command that is issued. For our purposes, we don't need that, but your use case might require it, for example, when a user decides to make a payment.

The fixed version is here, with the changes highlighted in bold:

```
openapi: 3.0.0
info:
  title: Home Automation API
  description: This API allows controlling lights and garage doors in
a home automation system.
  version: 1.0.0
servers:
  - url: https://skb-home-automation.azurewebsites.net
    description: Main server
paths:
  /operate_light:
    post:
      operationId: operateLight
      summary: Controls a light in the home automation system.
      x-openai-isConsequential: false
      requestBody:
        required: true
        content:
          application/json:
            schema:
              type: object
              required:
                - location
                - action
              properties:
                location:
                  type: string
                  description: The location of the light to be
controlled.
                  enum:
                    - "kitchen"
                    - "living room"
                    - "bedroom"
```

```
                    - "garage"

                action:
                  type: string
                  description: The action to be performed on the
light.
                  enum:
                    - "on"
                    - "off"
      responses:
        '200':
          description: Operation result
          content:
            application/json:
              schema:
                type: object
                properties:
                  result:
                    type: string
                    description: The result of the light operation.

  /operate_garage_door:
    post:
      operationId: operateGarageDoor
      summary: Controls the garage door in the home automation system.
      x-openai-isConsequential: false
      requestBody:
        required: true
        content:
          application/json:
            schema:
              type: object
              required:
                - action
              properties:
                action:
                  type: string
                  description: The action to be performed on the
garage door.
                  enum:
                    - "open"
```

```
                          - "close"

        responses:
          '200':
            description: Operation result
            content:
              application/json:
                schema:
                  type: object
                  properties:
                    result:
                      type: string
                      description: The result of the garage door
    operation.
```

You can paste this corrected version into the **Schema** box and click **Update** in the top-right corner. This will deploy the custom GPT with the connection to the native application you developed with Semantic Kernel.

Here, you will see an example of a real dialog with our custom GPT, where I ask it to operate several devices:

 You

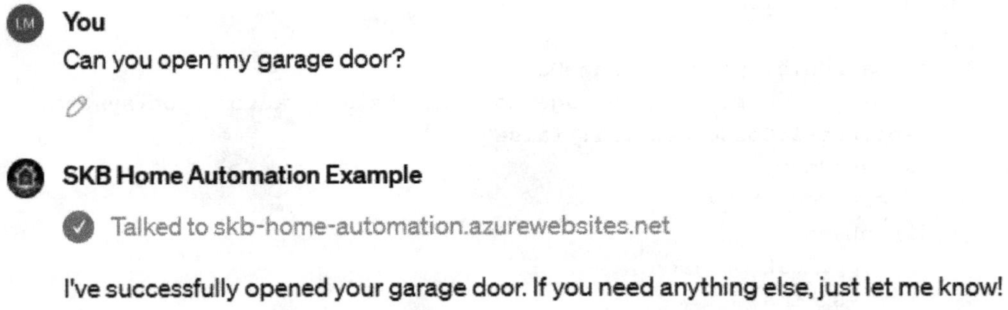

Can you open my garage door?

 SKB Home Automation Example

✓ Talked to skb-home-automation.azurewebsites.net

I've successfully opened your garage door. If you need anything else, just let me know!

Figure 8.11 – Using our SKB Home Automation custom GPT

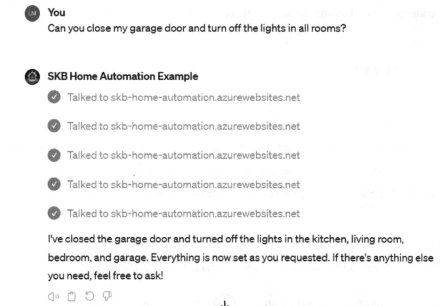

You

Can you close my garage door and turn off the lights in all rooms?

SKB Home Automation Example

✔ Talked to skb-home-automation.azurewebsites.net

✔ Talked to skb-home-automation.azurewebsites.net

✔ Talked to skb-home-automation.azurewebsites.net

✔ Talked to skb-home-automation.azurewebsites.net

✔ Talked to skb-home-automation.azurewebsites.net

I've closed the garage door and turned off the lights in the kitchen, living room, bedroom, and garage. Everything is now set as you requested. If there's anything else you need, feel free to ask!

Figure 8.12 – Using our SKB Home Automation custom GPT

First, I asked the GPT to open my garage door. It correctly called the API with the appropriate command. Next, I asked it a complex command: to close my garage door and turn off all the lights. As you can see from *Figure 8.12*, it issued five commands. Looking at the log of the web API, you would be able to see that the commands were correctly sent:

```
2024-04-08T05:16:27.968713802Z: [INFO]  Changed the status of the
garage door to close.
2024-04-08T05:16:30.939181143Z: [INFO]  Changed status of the kitchen
lights to off.
2024-04-08T05:16:33.701639742Z: [INFO]  Changed status of the living
room lights to off.
2024-04-08T05:16:36.377148658Z: [INFO]  Changed status of the bedroom
lights to off.
2024-04-08T05:16:39.017400267Z: [INFO]  Changed status of the garage
lights to off.
```

If I ask for a command that it cannot execute, it also correctly responds with what it can do:

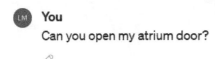

You
Can you open my atrium door?

SKB Home Automation Example
I'm only able to assist with the garage door and lights in specific rooms like the kitchen, living room, bedroom, and garage. Unfortunately, I can't help with the atrium door. If there's anything else within my capabilities that you need, just let me know!

Figure 8.13 – Issuing an invalid command to the custom GPT

The two main consequences of connecting an application with a custom GPT are as follows:

- **Your application gets a state-of-the-art planner**: Your users can make complex requests, such as issuing several commands at once, such as `turn off all my lights`, and ChatGPT will parse them and send them to your application. If users ask for functions that are not available in your application, ChatGPT tells them what can and cannot be done.

- **Your application gets wide distribution and access to all UI facilities provided by ChatGPT**: Everyone with access to ChatGPT can use your application, even from their phones. They can also use the application with their voices because ChatGPT supports voice commands.

In this section, we have seen how to connect an app we wrote with ChatGPT, enabling it to be used by the hundreds of millions of ChatGPT users.

Summary

In this chapter, we connected an application with OpenAI's ChatGPT by developing a custom GPT and adding custom actions to it. This can enable applications to get access to a planner that is based on the latest model available to ChatGPT users, which is usually a very advanced model.

In addition, what we have learned allows you to deploy your application to hundreds of millions of users with minimal effort and get access to several new features available to ChatGPT users, such as natural language requests and voice requests. It also allows you to deploy your application to users more quickly, as you don't have to develop a UI yourself – you can use ChatGPT as the UI as you develop and grow your application.

If you are a Python programmer, Microsoft Semantic Kernel provides a few additional features over what is already provided by the default OpenAI Python API. Among other things, you get the separation between prompt and code, native functions, planners, core plugins, and interfaces with memory. All of this can cut the time you will spend creating and maintaining your code. With the amount of change happening in AI these days, it's great to be able to save some time.

If you are a C# developer, in addition to getting the benefits that the Python programmers get, you will also find that Microsoft Semantic Kernel is the best way of connecting a C# application to OpenAI models, as OpenAI does not provide a C# API. You can do a lot with REST APIs, as we have shown when we created DALL-E 3 images in *Chapter 4*, but REST APIs are cumbersome and have changed in the last year. Using the Microsoft Semantic Kernel greatly simplifies things, and when changes happen, it's likely that they will be incorporated in a future release.

For now, this concludes our journey with Microsoft Semantic Kernel. As a parting thought, Semantic Kernel and AI models are just tools. Your impact on the world depends on how you use these tools. In my career, I have been fortunate to be able to use technology, and lately, AI, for social good. I hope you can do the same.

Index

packtpub.com

Subscribe to our online digital library for full access to over 7,000 books and videos, as well as industry leading tools to help you plan your personal development and advance your career. For more information, please visit our website.

Why subscribe?

- Spend less time learning and more time coding with practical eBooks and Videos from over 4,000 industry professionals
- Improve your learning with Skill Plans built especially for you
- Get a free eBook or video every month
- Fully searchable for easy access to vital information
- Copy and paste, print, and bookmark content

Did you know that Packt offers eBook versions of every book published, with PDF and ePub files available? You can upgrade to the eBook version at packtpub.com and as a print book customer, you are entitled to a discount on the eBook copy. Get in touch with us at customercare@packtpub.com for more details.

At www.packtpub.com, you can also read a collection of free technical articles, sign up for a range of free newsletters, and receive exclusive discounts and offers on Packt books and eBooks.

Other Books You May Enjoy

If you enjoyed this book, you may be interested in these other books by Packt:

OpenAI API Cookbook

Henry Habib

ISBN: 978-1-80512-135-0

- Grasp the fundamentals of the OpenAI API

- Navigate the API's capabilities and limitations of the API

- Set up the OpenAI API with step-by-step instructions, from obtaining your API key to making your first call

- Explore advanced features such as system messages, fine-tuning, and the effects of different parameters

- Integrate the OpenAI API into existing applications and workflows to enhance their functionality with AI

- Design and build applications that fully harness the power of ChatGPT

Modern Data Architecture on AWS

Behram Irani

ISBN: 978-1-80181-339-6

- Familiarize yourself with the building blocks of modern data architecture on AWS
- Discover how to create an end-to-end data platform on AWS
- Design data architectures for your own use cases using AWS services
- Ingest data from disparate sources into target data stores on AWS
- Build data pipelines, data sharing mechanisms, and data consumption patterns using AWS services
- Find out how to implement data governance using AWS services

Packt is searching for authors like you

If you're interested in becoming an author for Packt, please visit `authors.packtpub.com` and apply today. We have worked with thousands of developers and tech professionals, just like you, to help them share their insight with the global tech community. You can make a general application, apply for a specific hot topic that we are recruiting an author for, or submit your own idea.

Share Your Thoughts

Now you've finished *Building AI Applications with Microsoft Semantic Kernel*, we'd love to hear your thoughts! Scan the QR code below to go straight to the Amazon review page for this book and share your feedback or leave a review on the site that you purchased it from.

https://packt.link/r/1-835-46370-3

Your review is important to us and the tech community and will help us make sure we're delivering excellent quality content.

Download a free PDF copy of this book

Thanks for purchasing this book!

Do you like to read on the go but are unable to carry your print books everywhere?

Is your eBook purchase not compatible with the device of your choice?

Don't worry, now with every Packt book you get a DRM-free PDF version of that book at no cost.

Read anywhere, any place, on any device. Search, copy, and paste code from your favorite technical books directly into your application.

The perks don't stop there, you can get exclusive access to discounts, newsletters, and great free content in your inbox daily

Follow these simple steps to get the benefits:

1. Scan the QR code or visit the link below

https://packt.link/free-ebook/9781835463703

2. Submit your proof of purchase
3. That's it! We'll send your free PDF and other benefits to your email directly

www.ingramcontent.com/pod-product-compliance
Lightning Source LLC
Chambersburg PA
CBHW080636060326

40690CB00021B/4957